Glial Cell Engineering in Neural Regeneration

Li Yao

Glial Cell Engineering in Neural Regeneration

Springer

Li Yao
Department of Biological Sciences
Wichita State University
Wichita, KS, USA

ISBN 978-3-030-02103-0   ISBN 978-3-030-02104-7 (eBook)
https://doi.org/10.1007/978-3-030-02104-7

Library of Congress Control Number: 2018957838

© The Author(s) 2018

This work is subject to copyright. All rights are reserved by the Publisher, whether the whole or part of the material is concerned, specifically the rights of translation, reprinting, reuse of illustrations, recitation, broadcasting, reproduction on microfilms or in any other physical way, and transmission or information storage and retrieval, electronic adaptation, computer software, or by similar or dissimilar methodology now known or hereafter developed.

The use of general descriptive names, registered names, trademarks, service marks, etc. in this publication does not imply, even in the absence of a specific statement, that such names are exempt from the relevant protective laws and regulations and therefore free for general use.

The publisher, the authors, and the editors are safe to assume that the advice and information in this book are believed to be true and accurate at the date of publication. Neither the publisher nor the authors or the editors give a warranty, express or implied, with respect to the material contained herein or for any errors or omissions that may have been made. The publisher remains neutral with regard to jurisdictional claims in published maps and institutional affiliations.

This Springer imprint is published by the registered company Springer Nature Switzerland AG
The registered company address is: Gewerbestrasse 11, 6330 Cham, Switzerland

Preface

Neural tissue damage may result in lifelong disabilities and represents a considerable public burden. Advances in therapeutic approaches during the past few decades offer hope for such victims. However, the limited functional improvement in in vivo studies hinders effective application of therapeutic strategies in clinical practice due to the complexity of the nervous system. Very little spontaneous regeneration, repair, or healing occurs in the central nervous system. Neural tissue engineering that combines treatments of cell therapy, transplantation of biomaterial scaffolds, genetic manipulation, and electrical stimulation is one of the most promising methods to restore function of nervous system.

Glia play a pivotal role in maintaining the structural integrity and physiological function of the neurons. Glia are a fundamental cell type for the control of several critical activities of nervous system, such as myelination, synaptic transmission, and homeostasis. They are also involved in all types of pathological processes of the nervous system, including acute lesions such as trauma or stroke and chronic neurodegenerative disease. Glia transplantation may enhance functional recovery of a damaged nervous system by myelinating axons, providing trophic support, and promoting endogenous regeneration.

This book summarizes the advance of the research of glia function in neural regeneration and glia–biomaterials interaction. This book also reviews a unique function of electric fields in axonal growth, neural cell migration and division, and stem cell differentiation. The major contents of this book include the following: advances in the research of astrocyte function in neural regeneration (Chapter 1, by Madhulika Srikanth, Li Yao, and Ramazan Asmatulu); enhancement of axonal myelination in wounded spinal cord using oligodendrocyte precursor cell transplantation (Chapter 2, by Li Yao and Michael Skrebes); application of Schwann cells in neural tissue engineering (Chapter 3, by Li Yao and Priyanka Priyadarshani); stem cell- and biomaterial-based neural repair for enhancing spinal axonal regeneration (Chapter 4, by Pranita Kaphle, Li Yao, and Joshua Kehler); electric field-guided cell migration, polarization, and division: an emerging therapy in neural regeneration (Chapter 5, by Li Yao and Yongchao Li); vascularization in the spinal cord: the pathological process and therapeutic approach (Chapter 6, by Hien Tran and Li

Yao). I would like to acknowledge my laboratory members and colleagues for their contribution to this book.

Wichita, KS, USA　　　　　　　　　　　　　　　　　　　　　　　　　　Li Yao

Contents

1	**Advances in the Research of Astrocyte Function in Neural Regeneration**................................	1
	1.1 Astrocytes in the Central Nervous System	1
	1.2 Types of Astrocytes	2
	1.3 Functions of Astrocytes in CNS	3
	1.3.1 Control of Extracellular Homeostasis	3
	1.3.2 Removal of Excess Glutamate...................	4
	1.3.3 Maintenance of Glutamatergic Neurotransmission	4
	1.3.4 Control of Local Blood Flow and Metabolic Support for Neurons................	4
	1.3.5 Control of Synaptogenesis and Its Maintenance	4
	1.3.6 Tripartite Synapse	5
	1.3.7 Signaling in Glial Syncytia	5
	1.3.8 Concept of Gliotransmission	6
	1.4 Role of Astrocytes in Neurodegenerative Diseases.............	6
	1.5 Role of Astrocytes in Neural Injury.....................	7
	1.6 Importance of Astrocytes in Neural Regeneration	8
	1.7 Astrocyte Transplantation in Neural Regeneration	9
	References...................................	14
2	**Enhancement of Axonal Myelination in Wounded Spinal Cord Using Oligodendrocyte Precursor Cell Transplantation**.................................	19
	2.1 Axonal Myelination in Central Nervous System................	19
	2.2 Axon Demyelination Resulting from Spinal Cord Injury	20
	2.3 Restoration of Axonal Myelination Post-Spinal Cord Injury	21
	2.4 Transplantation of OPCs Enhancing Axon Remyelination in Spinal Cord Injury.......................	24
	2.4.1 Transplantation of OPCs in Therapy of SCI	24

		2.4.2 Co-Transplantation of Biomaterial Scaffolds and OPCs for SCI Therapy 25
	2.5	Potential Application of Electrical Stimulation in Axonal Myelination.......................... 28
	References.. 31

3 Application of Schwann Cells in Neural Tissue Engineering 37
  3.1 Origin of Schwann Cells 37
  3.2 Peripheral Nerve Injury and Role of Schwann Cells in Nerve Regeneration 38
  3.3 Autologous Nerve Grafting and Schwann Cell Transplantation for Nerve Repair........................ 40
  3.4 Synthetic and Biological Molecules for Promoting Nerve Repair................................. 44
  3.5 Stem Cell-Derived Schwann Cells for Neural Regeneration...................................... 45
  3.6 Schwann Cells and Biomaterial for Neural Regeneration...................................... 46
  3.7 Role of Schwann Cells in Spinal Cord Injury 50
  References.. 52

4 Stem Cell- and Biomaterial-Based Neural Repair for Enhancing Spinal Axonal Regeneration................ 59
  4.1 Stem Cell Therapy for Axonal Regeneration in Spinal Cord Repair 59
    4.1.1 Embryonic Stem Cells............................... 60
    4.1.2 Neural Stem Cells 66
    4.1.3 Induced Pluripotent Stem Cells....................... 68
    4.1.4 Mesenchymal Stem Cells............................ 69
  4.2 Biomaterial and Stem Co-Transplantation in Neural Regeneration 71
  References.. 75

5 Electric Field-Guided Cell Migration, Polarization, and Division: An Emerging Therapy in Neural Regeneration............................... 85
  5.1 Electric Field Directing Axonal Growth 85
  5.2 Electric Fields Directing Neuron Migration 88
  5.3 EF-Guided Migration of Stem Cells and Stem Cell–Derived Neural Cells................................ 89
  5.4 Oriented Cell Division in EFs 92
  5.5 Regulation of EF-directed Neuronal Migration.................. 93
    5.5.1 Cell Polarization in EFs.............................. 93
    5.5.2 Calcium Signals Growth Cone Navigation and Neuron Migration in Applied EF 96

Contents

        5.5.3   Cell Membrane Receptors and Intracellular
                    Signaling Pathways 96
   5.6   Electrical Activity in CNS Development
         and Regenerating Tissues.................................. 99
   5.7   Electric Fields Enhance Nerve and Spinal
         Cord Regeneration In Vivo 100
   5.8   Potential of Applying Electric Fields to Guide
         Cell Migration in Neurogenesis........................... 101
   References... 102

**6   Vascularization in the Spinal
Cord: The Pathological Process in Spinal
Cord Injury and Therapeutic Approach** 111
   6.1   Vascular Structure of Spinal Cord 111
   6.2   Damage of Vasculature in Spinal Cord Injury 113
   6.3   Natural Process of Revascularization
         and Remedies Post–SCI.................................. 114
   6.4   Therapeutic Approaches to Reconstruction
         of Vascular Structure Following SCI........................ 116
        6.4.1   Therapeutic Molecules............................. 116
        6.4.2   Biomaterial Scaffolds.............................. 120
        6.4.3   Extracorporeal Shock Wave Therapy.................. 121
        6.4.4   Hypothermia 122
   References... 123

Index 127

Chapter 1
Advances in the Research of Astrocyte Function in Neural Regeneration

Madhulika Srikanth, Li Yao, and Ramazan Asmatulu

Abstract Astrocytes play a critical role in maintaining the structural health of the neurons. Astrocytes create the brain environment by building up the micro-architecture of the central nervous system, maintain brain homeostasis, and control the metabolism of neural cells and synaptic activity. Astrocytes are involved in all types of brain pathologies from acute lesions to chronic neurodegenerative processes such as Alexander's disease, Alzheimer's disease, Parkinson's disease, multiple sclerosis, and psychiatric diseases. It was suggested that astrocytes play a negative role following the event of injury because they contribute to the formation of glial scar that inhibits the regeneration and growth of neurons. Recent compelling research shows that reactive astrocytes protect injured tissues and cells in various ways. Studies revealed that transplantation of astrocytes and glial-restricted precursor (GRP)-derived astrocytes (hGDAs) promoted neural regeneration process. In this chapter, we summarize the function of astrocytes in normal neural tissue and the cellular process of astrocytes in neural lesion. We also review the interaction of astrocytes and biomaterials and its potential application in neural regeneration.

Keywords Schwann cell · Neural regeneration · Myelination · Peripheral nerve · Axon · Biomaterials scaffolds · Gene therapy · Neural degeneration · Nerve injury · Transplantation

1.1 Astrocytes in the Central Nervous System

The central nervous system (CNS) consists of two main components: the brain and the spinal cord. It controls the physiological and psychological aspects of a living system and governs the biology within that system. Astrocytes are star-shaped glial cells that create the brain environment by building up the micro-architecture of the CNS. They have been recently established as one of the most important cells because they perform numerous vital functions. The death or survival of astrocytes affects the ultimate clinical outcome and rehabilitation through effects on neuron genesis and reorganization in the event of a trauma [1].

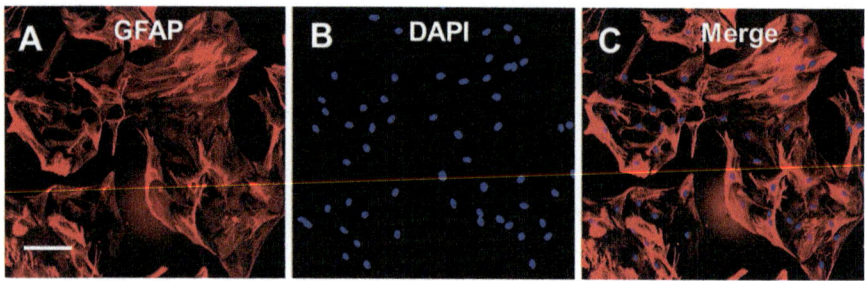

Fig. 1.1 Morphology of cultured astrocytes isolated from neonatal rat and GFAP expression: (**a**) Immunostaining of astrocytes with anti-GFAP antibody. (**b**) Nuclei of astrocytes labeled with DAPI. (**c**) Merging of images (**a**) and (**b**). Figure reproduced from Seyedhassantehrani et al. [8] with permission from Royal Society of Chemistry

About 90% of the human CNS consists of glial cells. Astrocytes which are a major type of glial cells play an essential role, both structurally and physiologically. A typical morphological feature of astrocytes is their expression of intermediate filaments that form the cytoskeleton. The main types of astroglial intermediate filament proteins are glial fibrillary acidic protein (GFAP) and vimentin. GFAP is commonly used as a specific marker for the identification of astrocytes (Fig. 1.1). They are very heterogeneous in nature. Each astrocyte extends processes that cover almost eight neuronal cell bodies, five blood vessels, and more than 100,000 synapses. Protoplasmic astrocytes are known to occupy a separate territory and create the micro-anatomical domains within the limits of their processes. Within these anatomical domains, the membrane of the astrocyte covers synapses and neuronal membranes. It also sends processes to cover the wall of the neighboring blood vessel with their end-feet. Astrocytes define the micro-architecture of the parenchyma by dividing the gray matter into relatively independent structural units in the mammalian brain. The astrocyte-neurons-blood vessel is known as a neurovascular unit [2–7].

1.2 Types of Astrocytes

Based on the morphology and immune-histological studies, there are two major types of astrocytes: Type 1 and type 2. This finding is based on studies of optic nerves in rats [9, 10].

Type 1 astrocytes are developed from the astrocytic lineage in white matter. They are characterized by the presence of numerous fibrils in their cytoplasm. Fibrous astrocytes are prevalent among myelinated axonal fibers mostly found in white matter tracts.

Type 2 astrocytes, which occur in gray matter, are known to be derived from bipotential progenitor cells. They have fewer fibrils within their cytoplasm compared with fibrous astrocytes. The cytoplasmic organelles are sparse in protoplasmic astrocytes. The astrocyte complexity, in particular within the protoplasmic subfam-

Table 1.1 Astrocyte cell markers [9]

	Type 1	Type 2
GFAP	Present	Present
A2B5	Absent	Present
Ran-2	Absent	Present
Tetanus toxin	Absent	Present

ily, increases along with phylogenic evolution. The bipotential progenitor cells can develop into either type 2 astrocytes or oligodendrocytes. Type 2 astrocytes are immunoreactive for both anti-GFAP and anti-A2B5 antibodies, which are known to be specific markers for immature oligodendrocytes (Table 1.1). The study of type 2 astrocytes is complicated since the cell marker is expressed for a short time in vitro and is dependent on the type of culture conditioning [9, 10]. Oligodendrocyte progenitors have been shown to exhibit multilineage competence. The cells may develop toward the phenotype of type-II astrocytes under certain environmental and genetic conditions [11]. However, the molecular mechanism that signals the development of oligodendrocyte or astrocyte phenotype has not been fully understood. One recent study revealed that the Hdac3 can directly activate Olig2 expression and therefore control oligodendrocyte-astrocyte fate decision. Hdac3 ablation resulted in a significant increase of astrocytes generation accompanied with a loss of oligodendrocytes [11].

Attempts have been made to study the surface properties of astrocytes. Research using atomic force microscopy (AFM) has revealed that astrocytes have irregular ridge-like structures that form a meshwork distributed throughout the surfaces of the cell body. Immunocytochemical studies show that the ridges are very fine bundles of tubulin and/or GFAP [10, 12].

1.3 Functions of Astrocytes in CNS

1.3.1 *Control of Extracellular Homeostasis*

Astrocytes participate in the control of extracellular homeostasis in the central nervous system. They regulate the concentrations of ions, neurotransmitters, and metabolites and regulate water movements via multiple molecular cascades. Neuronal activity leads to an increase in K^+ concentration from a resting level of about 3 mM to a maximum of 10–12 mM under physiological conditions and to higher values under pathological conditions. A higher K^+ concentration in the extracellular space modulates neuronal activity. Astrocytes remove excess extracellular K^+ to maintain stability. Glial syncytia and aquaporin channels expressed in astrocytes play a role in water homeostasis in the brain [2, 4, 7, 13–17].

1.3.2 Removal of Excess Glutamate

Glutamate is the major excitatory neurotransmitter in the brain of all vertebrates. Excess glutamate acts as a potent neurotoxin, triggering neuronal cell death in acute and chronic brain lesions. Astrocytes remove the bulk of glutamate from the extracellular space. They can accumulate almost 80% of the glutamate that is released. Glutamate is removed by excitatory amino acid transporters (EAATs). Five types of EAATs are present in the human brain, but only two are expressed exclusively in astrocytes [18, 19].

1.3.3 Maintenance of Glutamatergic Neurotransmission

Astroglial glutamate transport is crucial for neuronal glutamatergic transmission by operating the glutamate-glutamine pathway. Glutamate that is accumulated by astrocytes is converted into glutamine. Glutamine is not sensed by neurotransmitter receptors and is not toxic. It can be safely transported to presynaptic terminals from the extracellular space and converted back to glutamate [2].

1.3.4 Control of Local Blood Flow and Metabolic Support for Neurons

Astrocytes integrate neural circuitry with local blood flow and metabolic support. The basal lamina of blood vessels is almost entirely covered by astrocyte end-feet. One arm reaches the blood vessel and the other attaches to the neuronal membrane, synapse, or axon. Increased neuron activity triggers Ca^{2+} signals in astrocytes, which leads to the release of vasoactive agents that regulate the local blood flow. It is still unknown if astrocyte activity leads to vasoconstriction or vasodilatation. However, astrocytes are the only cells in the brain that can synthesize glycogen and thus serve as an energy reservoir [2, 20–23].

1.3.5 Control of Synaptogenesis and Its Maintenance

Astrocytes regulate the formation, maturation, maintenance, and stability of synapses and hence control the connectivity of neuronal circuits. They secrete numerous chemical factors required for synaptogenesis. Synaptic formation depends on the formation of cholesterol, which is produced and secreted by astrocytes. This cholesterol can be converted into steroid hormones that can act as synaptogenic signals. It is also induced by the expression of specific proteins, agrin and thrombin, which are essential for synapse formation. Subsequently, astrocytes control the maturation of synapses through several signaling systems, which affect the postsynaptic density and synapse maturation such as tumor necrosis factor alpha and the activity-dependent neurotrophic factor. Astrocytes can limit the number of synapses since their membranes can enclose the neuronal processes and thus compete with synapses. They can also eliminate synapses in the CNS by secreting certain factors or proteolytic enzymes, which demolish the extracellular matrix (ECM) and reduce the stability of the synaptic contact [2, 21, 23–25].

1.3.6 Tripartite Synapse

In the gray matter, astrocytes are associated with neuronal membranes and synaptic regions, so that astroglial membranes completely or partially enwrap presynaptic terminals and postsynaptic structures. In the cerebellum, glial-synaptic relations are even closer, and almost all of the synapses formed by parallel fibers on the dendrites are covered by the membranes of Bergmann glial cells, which enwrap between 2000 and 6000 synaptic contacts. The distance between terminal structures of astrocytes and neuronal presynaptic and postsynaptic membranes is almost 1 µm and therefore is exposed to the neurotransmitters released from the synaptic terminals. They have a complement of receptors very similar to that of their neuronal neighbor. In the cortex, both pyramidal neurons and neighboring astroglial cells express glutamate and purinoreceptors, whereas in the basal ganglia neurons and astrocytes are sensitive to dopamine. The astroglial cell closely resembles the postsynaptic neuron [2].

Close morphological and functional relations between astrocytes and synapses lead to the concept of the "tripartite synapse," whereby synapses are built from three equally important parts: The presynaptic terminal, the postsynaptic neuronal membrane, and the surrounding astrocyte. A neurotransmitter that is released from the presynaptic terminal activates receptors in both the postsynaptic neuronal membrane and the perisynaptic astroglial membranes, which results in the generation of a postsynaptic potential in the neuron and a Ca^{2+} signal in the astrocyte. It is difficult to know if astrocytes actively participate in the ongoing synaptic transmission because their signals are on a much slower time scale compared to the rapid signaling of neurons. Therefore, they have been considered as integrators or modulators [2, 5, 26–31].

1.3.7 Signaling in Glial Syncytia

Astroglial metabotropic receptors are coupled to intracellular signaling cascades, which provide glia with specific excitability mechanisms. This glial activity is based on the excitability of the endoplasmic reticulum membrane containing Ca^{2+} release channels. Stimulation of astroglial metabotropic receptors induces the formation of

InsP$_3$, which in turn triggers Ca^{2+} to be released from the endoplasmic reticulum, thus producing Ca^{2+} signals. These signals can cross cell-to-cell boundaries and propagate through astroglial syncytia. Mechanisms of generation and the maintenance of intercellular Ca^{2+} waves are complex [2].

1.3.8 Concept of Gliotransmission

Astrocytes can release a variety of transmitters into the extracellular space, such as glutamate, adenosine triphosphate (ATP), gamma-aminobutyric acid (GABA), and D-serine. Their mechanisms include diffusion through high-permeability channels (Cl$^-$ channels) or through transporters or Ca^{2+}-dependent exocytosis [2, 28, 31–34].

1.4 Role of Astrocytes in Neurodegenerative Diseases

The pathological potential of neuroglia was recognized at the end of the nineteenth to the beginning of the twentieth centuries. Astrocytes are an important part of the intrinsic brain defense system. Brain insults can trigger an evolutionary conserved astroglial defense response generally referred to as reactive astrogliosis, which is essential for both limiting the area of damage and for post-insult remodeling and recovery of neural function [7, 35–37].

Astrocytes are involved in all types of brain pathologies from acute lesions (trauma or stroke) to chronic neurodegenerative processes (Alexander's disease, Alzheimer's disease [AD], Parkinson's disease, multiple sclerosis, and many others) and psychiatric diseases. Recent studies have highlighted the role of astroglial degeneration and atrophy in the early stages of various neurodegenerative disorders, which may be important in cognitive impairment. Collectively, astrocytes play an important role in the progression and outcome of neurological diseases [2].

It has been reported that astrocytes react to various neurodegenerative insults rapidly, leading to vigorous astrogliosis. This reactive gliosis is associated with alteration in morphology and structure of activated astrocytes along with its functional characteristics [38]. The astrocytic processes construct a bushy network surrounding the injury site, thus secluding the affected part from the rest of the central nervous system. Subsequently, astrogliosis has been implicated in the pathogenesis of a variety of chronic neurodegenerative diseases, including AD, Parkinson's disease, amyotrophic lateral sclerosis (ALS), acute traumatic brain injury, stroke, and neuroinflammatory brain diseases.

One of the trademarks of various neurodegenerative diseases and neuroinflammatory disorders is oxidative stress-induced CNS damage. Similarly, the natural aging process is associated with increased oxidative stress. Such oxidative stress can damage lipids, proteins, and nucleic acids of cells and power-house mitochondria causing cell death in assorted cell types including astrocytes and neurons.

Mitochondria are central neuronal organelles that play a vital role in neuronal life and death. Both mitochondrial dysfunction and proper function are essential components in neurodegeneration [39].

Some neurodegenerative diseases like ALS lead to the loss of motor neurons due to mutations of the superoxide dismutase (SOD1) enzyme. It has been shown that the SOD1 gene knockout mice diminished the mutant expression in astrocytes, delayed microglial activation, and sharply slowed the disease progression. Research has shown that mutant astrocytes are viable targets for therapies for slowing the progression of non-cell autonomous killing of motor neurons in ALS [40].

1.5 Role of Astrocytes in Neural Injury

The role of astrocytes after a neural injury has been misunderstood for several years. It is known that astrocytes play a negative role following the event of injury that inhibits the regeneration and growth of neurons [41]. Recent studies indicate findings that contribute to the understanding of astrocyte function in neural regeneration. Cellular damage can result in the release of large amounts of ATP into the extracellular environment, because intracellular concentrations can reach 3–5 mM [42–44]. TNFα directly affects astrocytes by inducing a slow increase in intracellular Ca^{2+} and marked depolarization with the consequence of disturbing voltage-dependent glial functions. This finding suggested that the immuno-electrical coupling may contribute to an impairment of neuronal function in inflammatory CNS diseases [45]. Evidence shows that the interastrocytic gap junctional communication decreases the vulnerability of the neurons to oxidative injury by a process of cellular calcium homeostasis and dissipation of oxidative stresses [46]. Pro-inflammatory cytokines, such as IL-1 and TNFα, can cause damage to the blood-brain barrier, release neurotoxins such as nitric oxide from the vascular endothelium, cause the up-regulation of adhesion molecules involved in the invasion of leukocytes (for example, intercellular cell-adhesion molecule 1), and induce vasogenic edema [44]. Nucleosides and nucleotides released from dying cells might induce reactive astrogliosis, which involves striking changes in astrocyte proliferation and morphology [41, 42, 44, 47]. This can be measured by increased expression of the astroglial-specific marker GFAP and the elongation of GFAP-positive processes [48]. ATP stimulates the proliferation of microglia and acts as a powerful chemoattractant to the site of brain injury [49, 50]. In moderate situations (chemical lesions and mild injury), reversible aberrant hypertrophy of astrocytes occurs. This is marked by up-regulated GFAP without any significant proliferation [7, 51–53]. Over time, the severity of injury increases with the expansion of astrocytes both in terms of hypertrophy and proliferation, which gradually disrupt and distort the once well-organized tissues [7]. This may lead to irreversible glial scarring, which is predominantly composed of reactive astrocytes, resident microglia, infiltrating macrophages, and ECM molecules. This feature has been thought to inhibit repair and to be associated with dystrophic axons [54, 55].

Recent compelling research shows that reactive astrocytes protect injured tissues and cells in various ways [35, 56]. One major benefit brought by reactive astrocytes is neuroprotection that is mediated by the degradation of amyloid-beta peptides [57] and the release of neurotrophic factors and growth supportive factors, such as brain-derived neurotrophic factor (BDNF) [58], ciliary neurotrophic factor (CNTF), and laminin [59]. By clearing excitotoxic glutamate with increased expression of glial transporters such as the Glu-Asp transporter (GLAST) and glial glutamate transporter 1 (GLT-1), reactive astrocytes protect the spared neurons from oxidative stress and NH4+ toxicity [60–63]. Moreover, selective proliferation of juxtavascular astrocytes repairs the integrity of the blood-brain barrier (BBB) and blood spinal barrier (BSB) [60], and restricts the spread of inflammation [64]. In addition, astrocytes also stabilize the extracellular fluid and ion balance and reduce vasogenic edema after trauma. Hence, the sole ablation of activated astrocytes could lead to impaired restoration of the BSB, increased demyelination, enhanced inflammatory response, severe neuronal death, and thereafter worse locomotion recovery after spinal cord injury (SCI) [54, 60, 64, 65].

Glial scarring takes a considerable time to form a growth-blocking barrier, so it may not be the main reason for the poor regeneration of the central nervous system. It is gradually being recognized that glial scarring is more likely to be a required physiological response for restoring the internal homeostasis of the CNS that allows the cells to be replaced and the BSB to be restored by separating the remaining healthy tissue from the bleeding and necrotic primary lesion. This prevents the potential exacerbation of an inflammatory response, cellular death, and tissue damage during the secondary injury [35, 41, 54].

There are multiple unanswered questions about the source of astrocytes during astrogliosis. Studies indicate that proliferating cells can arise from either mature astrocytes or precursors and from either fibrous or protoplasmic astrocytes. There might be differences in the ability of protoplasmic and fibrous or type 1 and type 2 astrocytes to contribute to these lesions. Additionally, type 1 astrocytes are believed to be the primary cells responsible for formation of a glial scar in white matter areas and type 2 in gray matter areas [7, 10, 53, 66–68].

1.6 Importance of Astrocytes in Neural Regeneration

A previous study has shown that immature astrocytes were able to support robust neurite outgrowth and reduce scarring after brain injury. When transplanted into injured brain, the immature, but not mature, astrocytes had a limited ability to form bridges across the inhibitoriest outer rim. In turn, the astrocyte bridges could promote adult sensory axon regrowth. The use of selective enzyme inhibitors revealed that matrix metalloproteinase-2 (MMP-2) enabled immature astrocytes to cross the proteoglycan rim. In another study, when faced with a minimal lesion, neurons of the basal forebrain can regenerate in the presence of a proper bridge across the lesion and when levels of chondroitin sulfate proteoglycans (CSPGs) in the glial scar are reduced [69].

During brain ischemia, impairment in astrocyte functions can critically influence neuron survival. Astrocyte functions that are known to influence neuronal survival include glutamate uptake, glutamate release, free radical scavenging, water transport, and production of cytokines and nitric oxide. Astrocyte surface molecule expression and trophic factor release influence long-term recovery after brain injury. This is because their behavior affects neurite outgrowth, synaptic plasticity, and neuron regeneration. The death or survival of astrocytes themselves may affect the ultimate clinical outcome and rehabilitation through effects on neurogenesis and synaptic reorganization [70].

Astroglial cells are engaged in neurological diseases by determining the progression and outcome of neuropathological process. Astrocytes have been shown to release apolipoprotein E (ApoE), which has been shown to regulate neurotransmission, growth factor release, and immune responses [71]; therefore, they are involved in AD, ALS, Parkinson's disease, and various forms of dementia. Recent research findings have shown that early stages of neurodegenerative processes are associated with the atrophy of astroglia, which causes disruptions in synaptic connectivity, misbalance in neurotransmitter homeostasis, and neuronal death through increased excitotoxicity. In later stages, astrocytes become activated and contribute to the neuroinflammatory component of neurodegeneration [72].

In another study, Cheng and Kisaalita used three types of scaffolds—nano, micro, and nanofiber micropore (NFMP)—to study the differentiation of neural stem cells (NSCs). The microscaffolds and NFMP scaffolds exhibited significantly larger cell populations than the nanoscaffolds, probably due to the fact that the micropores within the micro- and NFMP scaffolds provided space for cells to infiltrate and grow in three dimensions. In contrast, the nanopores in the nanoscaffolds were too small for cells to infiltrate; therefore, the cells could only grow on the surface. The nanostructures within NFMP scaffolds induced partial neural differentiation and suppressed cell proliferation. To quantify the extent of neural differentiation, two intracellular markers—beta-tubulin III (Tuj) for neurons and nestin for neural progenitors—were used. A large number of cells from all three groups stained positive for both Tuj (red) and nestin (green) at 14 days into differentiation. According to the quantitative differentiation, the microscaffolds exhibited a significantly smaller percentage of Tuj positive cells, compared to the nano- and NFMP scaffolds, both of which were undistinguishable from each other with regard to the Tuj positive cell percentage. Moreover, Tuj positive neurites, which were four times longer than the soma bodies, were found in nano- and NFMP scaffolds, while cells on the microscaffolds only developed short neurites [73].

1.7 Astrocyte Transplantation in Neural Regeneration

Different types of glial cells have been studied in cell transplantation therapy including rodent immature astrocytes and glial-restricted precursor (GRP)-derived astrocytes (hGDAs) with specifically identified phenotypes. In most of the transplantation methods, the astrocytes are locally injected into the injured site at different times after neural injury. One of the earlier forms of astrocyte transplantation was done in 1990 by Kliot et al. [74]. Here, embryonic astrocytes from the spinal cord were locally injected 7 days after spinal cord injury, which led to the growth of injured dorsal root fibers with long-lasting terminals formed by regenerated fibers. The transplantation generated limited inflammatory response. In another study, astrocytes isolated from the cerebral cortex of a new born rat were transplanted into a hemisected adult rat spinal cord. This study showed that the transplantation of astrocytes reduced the scar tissue formation [75]. In another study, neonatal astrocytes were locally injected into the hemisected spinal cord of Wistar rats, which were able to achieve subtle motor function improvement of the hind legs [76]. The Pröschel Laboratory generated enriched populations of two distinct types of astrocytes from GRP cells. These astrocyte populations, referred to as GRP-derived astrocyte-bone morphogenetic protein (BMP and astrocyte-CNTF (GDA^{BMP} and GDA^{CNTF}, respectively), differ in several properties, including their morphology, marker expression, and ability to support neurite outgrowth. GDA^{BMP} has been likened to astrocytes found during normal development, while GDA^{CNTF} resembles reactive astrocytes formed in response to numerous CNS insults. According to their research, hGDA^{BMP} promotes neuronal survival, axonal growth, and recovery of volitional foot placement, while hGDA^{CNTF} does not [77, 78]. Liu et al. reported numerous factors that can influence the efficiency of stem cell transplantation therapy. Astrocytes can secrete neurotrophins, chemokines, and cytokines under different conditions, through which it can regulate the proliferation, differentiation, and migration of neural stem cells [79].

Astrocyte transplantation has been studied by several research groups for stroke recovery. The mechanisms related to recovery from stroke are based on structural and functional changes in brain circuits that mainly take place in unaffected parts of the brain, which is insufficient to promote recovery of neurological functions. Glial cells that are reputed to remove excitatory neurotransmitters and electrolytes from the extracellular space, thus controlling neuronal excitability and enabling synaptic plasticity, become dysfunctional. Plasticity processes can inhibit tissue damage and loss of function after a stroke. Research conducted by Luo et al. shows that co-transplantation of astrocytes and NSCs into the ischemic striatum of transient middle cerebral artery occlusion (MCAO) model rats resulted in a higher ratio of survival and proliferation of the transplanted NSCs, and neuronal differentiation [80].

In cases with middle cerebral artery occlusion, a delayed intravenous transplantation of neural precursor cells in mice can reduce ischemia-induced changes in the ipsilesional hemisphere and promote contralesional adaptive plasticity. These cells are selectively located within the lesion and within perilesional brain areas and help to tone down excitatory neuronal networks by reducing the extracellular glutamate [81]. During cerebral ischemia, the excitotoxicity release of free radicals and the exacerbated immune response cause serious complications in motor and cognitive areas during post-ischemia. Cyclin-dependent kinase 5 (CDK5) is widely involved in the functions of neurons and astrocytes, and its overactivation is implicated in neurodegenerative processes. Some researchers have evaluated the brain parenchymal response to the transplantation of CDK5-killing astrocytes into the somatosensory cortex, and showed motor and neurological function recovery in rats [82].

In the field of ALS, where directly targeting the spinal cord is challenging and an invasive application, intraparenchymal injections of mesenchymal stem cells and neural stem cells into the thoracolumbar region, and of GRPs into the cervical region have been reported to demonstrate success in animal models in several research studies [83–86]. Research conducted by Li et al. evaluated intraspinal transplantation of human cell-derived astrocytes following cervical contusion SCI as an innovative strategy for reconstituting GLT-1 function and for protecting neurons related to diaphragmatic respiration. Astrocytes express the major glutamate transporter, GLT-1, which is responsible for the majority of glutamate uptake in the spinal cord. Following SCI, compromised GLT-1 function can increase susceptibility to excitotoxicity. According to their studies, transplant-derived cells showed robust long-term survival post-injection and efficiently differentiated into astrocytes in injured spinal cord of both immune-suppressed mice and rats [87].

Transplantation of wild-type astrocytes or their precursors into CNS tissue affected by ALS is another promising therapeutic approach, with the aim of rescuing motor neurons responsible for breathing, which is the primary cause of death in human ALS. In research conducted by Lepore et al., GRPs survived in diseased tissue that differentiated efficiently into astrocytes, and reduced microgliosis in the cervical spinal cord. GRPs extended survival and disease duration, attenuated motor neuron loss, and slowed declines in forelimb motor and respiratory function [85, 88]. ALS and spinal cord injury affect long segments of the spinal cord that need multiple treatments, which adds to the complexity of the treatment. Neurosurgeons are developing new surgical techniques or tools to deliver therapeutic agents to large CNS areas with a single treatment using different types of delivery routes like intranasal, intrathecal, intraperitoneal, intramuscular, intravenous, and intra-bone marrow cell transplantation [88–93].

Natural biomaterials such as fibrin and collagen hydrogels act as a growth-permissive carrier for cell transplantation into an injured spinal cord. However, the relatively rapid degradation of collagen and fibrin limits their application. In one study, fibrin and collagen hydrogels were chemically treated with aprotinin and poly(ethylene glycol) ether tetrasuccinimidyl glutarate (4S-StarPEG), respectively, to reduce the cell-mediated degradation rate [8]. The immature astrocytes isolated from neonatal rats were grown into the hydrogels and the cell behavior studied. The cell viability assay showed that astrocytes maintained good viability in these hydrogels. The cell motility of astrocytes in these hydrogels was analyzed using a time-lapse imaging system, and the migration speed was quantified. The cross-linking of collagen hydrogel with 4S-StarPEG did not change the astrocyte migration speed. However, the addition of aprotinin in the fibrin hydrogel inhibited astrocyte migration (Fig. 1.2).

The microenvironment of the neural tissue lesion that the transplanted cells encountered is unfavorable for cell survival. A cell-delivery vehicle such as microspheres may generate a permissive environment for the growth of grafted cells. Microspheres are a promising carrier for cell delivery into injured neural tissue for regeneration. In a previous study, the collagen microspheres encapsulating astro-

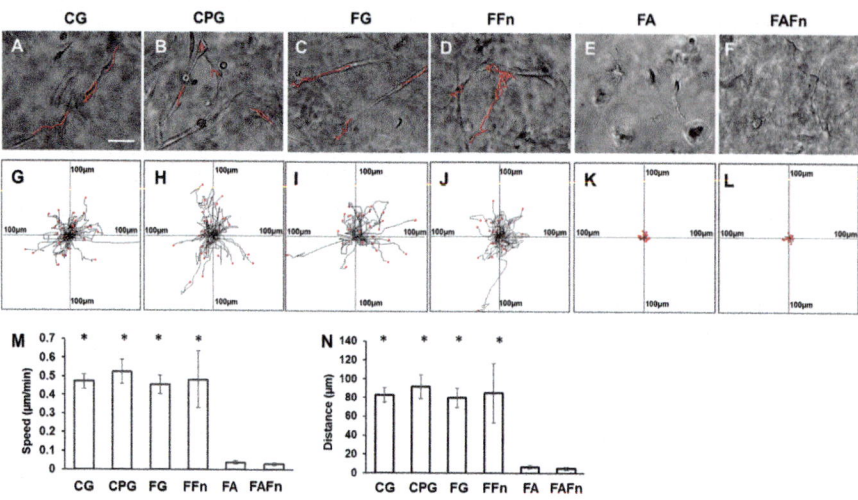

Fig. 1.2 Migration of astrocytes in hydrogels: (**a–f**) Astrocytes migrated in hydrogels (red lines indicate tracks of migration of astrocytes in hydrogels). Scale bar: 50 μm. (**g–l**) Cell migration paths determined by video monitor tracings (position of all cells at t = 0 min represented by origin position (center of frame), with migratory track of each cell at 3 h plotted as single line on graph; each arm of axes represents 100 μm of translocation distance). (**a, g**) Collagen hydrogel without cross-linking. (**b, h**) Collagen hydrogel cross-linked with 4S-StarPEG (0.05 mM). (**c, i**) Fibrin hydrogel without aprotinin. (**d, j**) Fibrin hydrogel with fibronectin (25 μg/ml). (**e, k**) Fibrin hydrogel with aprotinin (40 μg/ml). (**f, l**) Fibrin hydrogel with fibronectin (25 μg/ml) and aprotinin (40 μg/ml). (**m**) Quantification of cell migration speed in hydrogels. (**n**) Quantification of cell migration distance in hydrogels. $*p < 0.05$, compared with astrocyte migration in FA or FAFn, *CG* collagen hydrogel, *CPG* cross-linked collagen hydrogel, *FG* fibrin hydrogel, *FFn* fibrin hydrogel containing fibronectin, *FA* fibrin hydrogel containing aprotinin, *FAFn* fibrin hydrogel containing aprotinin and fibronectin. Figure reproduced from Seyedhassantehrani et al. [8] with permission from Royal Society of Chemistry

cytes were fabricated by injecting a mixture of collagen and astrocytes into a cell culture medium [94]. The size of the microspheres can be controlled by regulating the flow rate of the collagen solution in a syringe attached to a syringe pump. Cross-linking of collagen microspheres with 4S-StarPEG reduced the degradation rate of the microspheres against collagenase digestion. Viability of the cells in the cross-linked microspheres was higher than 90%. Astrocytes transfected with plasmids encoding nerve growth factor (NGF)-ires-enhanced green fluorescent protein (EGFP) genes were encapsulated in collagen microspheres. The NGF level generated by the astrocytes in the microspheres was determined by the enzyme-linked immunosorbent (ELISA) assay. The level of NGF released into the cell culture medium was higher than that remaining in the microsphere or in the astrocytes. The NGF generated by the astrocytes significantly enhanced the axonal growth of cultured rat dorsal root ganglion. This study suggested that collagen microspheres can efficiently encapsulate astrocytes and that the astrocytes delivered by the microspheres can potentially be used to promote neural regeneration (Fig. 1.3).

1.7 Astrocyte Transplantation in Neural Regeneration

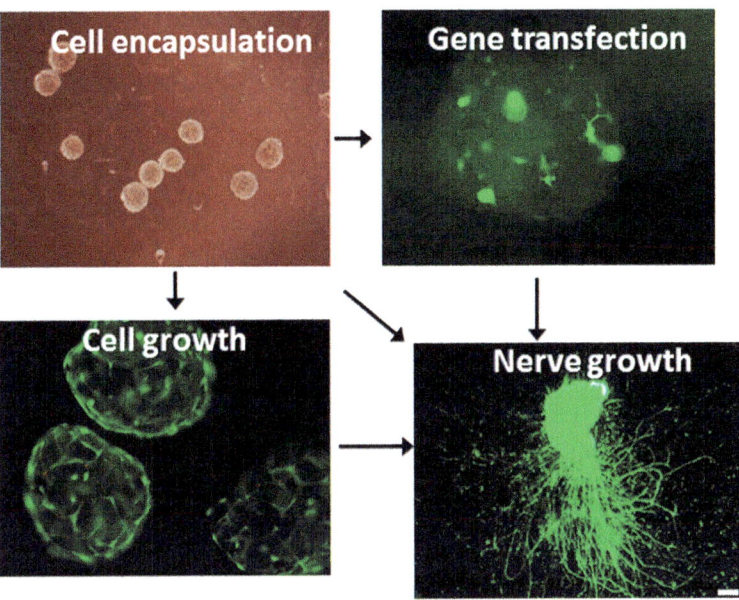

Fig. 1.3 Improvement of dorsal root ganglion axonal growth using microspheres encapsulating cells transfected with therapeutic gene. Figure reproduced from Berndt et al. [94] with permission from American Chemical Society

One study introduced a method to create a tubular neural tissue construct containing unidirectional neuron bundles using a scaffold-free approach. The researchers fabricated a neural tissue construct containing unidirectional neuron bundles based on surface patterning of a thermos-responsive culture substrate and a co-culture system of neurons with patterned astrocytes without the use of scaffolds. Surface patterning was able to provide an anisotropic structure. The neuron bundles can be laid out in the same direction at regulated intervals within multilayered astrocyte sheets by including a gelatin gel-coated plunger [95]. Another research study established the potential utility of astrocyte alignment. A three-dimensional astrocyte-seeded collagen gel culture system was used to explore the effect of astrocyte alignment on neuronal growth. The plastic compressed aligned gels can be used as mechanically robust implantable devices for neural regeneration [96].

Since, the importance of astrocytes has been realized only in the last a couple of decades, their functions and mechanism are not yet completely understood. Nevertheless, considerable progress has been made and reported by various researchers. Extensive research is necessary to define astrocyte intracellular signaling pathways that drive reactive changes in gene expression and cell function. This is a new area for identifying potential therapeutic targets for neurological disorders. The study of transcriptional regulatory molecules that ultimately transmit extracellular stimuli into changes in gene expression positions them as promising targets to selectively manipulate subtypes of reactive astrocytes that play a role across a broad range of central nervous system disorders [97, 98].

An efficient delivery method is still not completely developed and is a matter of deliberation. Several factors remain critical issues, including the dose of transplanted cells, timing, location, type of cells for particular needs, and cell migratory abilities. All of these must be taken into account for designing CNS stem cell-based therapies. Before broad clinical application, numerous preclinical studies using animal transplantation paradigms will be required to assess biodistribution, viability, integration into host tissue, differentiation into functional cells, lack of tumorigenicity, and safety of delivery [99–101].

References

1. Molofsky AV, Krencik R, Ullian EM, Tsai HH, Deneen B, Richardson WD, et al. Astrocytes and disease: a neurodevelopmental perspective. Genes Dev. 2012;26(9):891–907.
2. Kettenmann H, Verkhratsky A. Neuroglia - living nerve glue. Fortschr Neurol Psyc. 2011;79(10):588–97.
3. Allen NJ, Barres BA. Neuroscience: glia - more than just brain glue. Nature. 2009;457(7230):675–7.
4. Kimelberg HK, Nedergaard M. Functions of astrocytes and their potential as therapeutic targets. Neurotherapeutics. 2010;7(4):338–53.
5. Nedergaard M, Ransom B, Goldman SA. New roles for astrocytes: redefining the functional architecture of the brain. Trends Neurosci. 2003;26(10):523–30.
6. Oberheim NA, Wang X, Goldman S, Nedergaard M. Astrocytic complexity distinguishes the human brain. Trends Neurosci. 2006;29(10):547–53.
7. Sofroniew MV. Reactive astrocytes in neural repair and protection. Neuroscientist. 2005;11(5):400–7.
8. Seyedhassantehrani N, Li Y, Yao L. Dynamic behaviors of astrocytes in chemically modified fibrin and collagen hydrogels. Integ Biol. 2016;8(5):624–34.
9. Raff M, Abney E, Cohen J, Lindsay R, Noble M. Two types of astrocytes in cultures of developing rat white matter: differences in morphology, surface gangliosides, and growth characteristics. J Neurosci. 1983;3(6):1289–300.
10. Montgomery D. Astrocytes: form, functions, and roles in disease. Vet Pathol. 1994;31(2):145–67.
11. Zhang L, He X, Liu L, Jiang M, Zhao C, Wang H, et al. Hdac3 interaction with p300 histone acetyltransferase regulates the oligodendrocyte and astrocyte lineage fate switch. Dev Cell. 2016;36(3):316–30.
12. Yamane Y, Hatakeyama D, Haga H, Abe K, Ito E. Surface structures of cultured type 2 astrocytes revealed by atomic force microscopy. Zool Sci. 1999;16(1):1–7.
13. Dong Y, Benveniste EN. Immune function of astrocytes. Glia. 2001;36(2):180–90.
14. Nicoll JA, Weller RO. A new role for astrocytes: beta-amyloid homeostasis and degradation. Trends Mol Med. 2003;9(7):281–2.
15. Noble M, Fok-Seang J, Cohen J. Glia are a unique substrate for the in vitro growth of central nervous system neurons. J Neurosci. 1984;4(7):1892–903.
16. Fallon JR. Preferential outgrowth of central nervous system neurites on astrocytes and Schwann cells as compared with nonglial cells in vitro. J Cell Biol. 1985;100(1):198–207.
17. Obara M, Szeliga M, Albrecht J. Regulation of pH in the mammalian central nervous system under normal and pathological conditions: facts and hypotheses. Neurochem Int. 2008;52(6):905–19.
18. Sattler R, Rothstein JD. Regulation and dysregulation of glutamate transporters. Handb Exp Pharmacol. 2006;175:277–303.

References

19. Seifert G, Schilling K, Steinhauser C. Astrocyte dysfunction in neurological disorders: a molecular perspective. Nat Rev Neurosci. 2006;7(3):194–206.
20. Ballabh P, Braun A, Nedergaard M. The blood-brain barrier: an overview: structure, regulation, and clinical implications. Neurobiol Dis. 2004;16(1):1–13.
21. Gordon GR, Mulligan SJ, MacVicar BA. Astrocyte control of the cerebrovasculature. Glia. 2007;55(12):1214–21.
22. Haseloff RF, Blasig IE, Bauer HC, Bauer H. In search of the astrocytic factor(s) modulating blood-brain barrier functions in brain capillary endothelial cells in vitro. Cell Mol Neurobiol. 2005;25(1):25–39.
23. Koehler RC, Gebremedhin D, Harder DR. Role of astrocytes in cerebrovascular regulation. J Appl Physiol. 2006;100(1):307–17.
24. Qian XM, Shen Q, Goderie SK, He WL, Capela A, Davis AA, et al. Timing of CNS cell generation: a programmed sequence of neuron and glial cell production from isolated murine cortical stem cells. Neuron. 2000;28(1):69–80.
25. Schummers J, Yu HB, Sur M. Tuned responses of astrocytes and their influence on hemodynamic signals in the visual cortex. Science. 2008;320(5883):1638–43.
26. Oliet SH, Piet R, Poulain DA. Control of glutamate clearance and synaptic efficacy by glial coverage of neurons. Science. 2001;292(5518):923–6.
27. Fellin T. Communication between neurons and astrocytes: relevance to the modulation of synaptic and network activity. J Neurochem. 2009;108(3):533–44.
28. Halassa MM, Fellin T, Takano H, Dong JH, Haydon PG. Synaptic islands defined by the territory of a single astrocyte. J Neurosci. 2007;27(24):6473–7.
29. Fiacco TA, McCarthy KD. Astrocyte calcium elevations: properties, propagation, and effects on brain signaling. Glia. 2006;54(7):676–90.
30. Haydon PG, Carmignoto G. Astrocyte control of synaptic transmission and neurovascular coupling. Physiol Rev. 2006;86(3):1009–31.
31. Perea G, Araque A. Synaptic information processing by astrocytes. J Physiology-Paris. 2006;99(2–3):92–7.
32. Shigetomi E, Bowser DN, Sofroniew MV, Khakh BS. Two forms of astrocyte calcium excitability have distinct effects on NMDA receptor-mediated slow inward currents in pyramidal neurons. J Neurosci. 2008;28(26):6659–63.
33. Ullian EM, Sapperstein SK, Christopherson KS, Barres BA. Control of synapse number by glia. Science. 2001;291(5504):657–61.
34. Barres BA. The mystery and magic of glia: a perspective on their roles in health and disease. Neuron. 2008;60(3):430–40.
35. Silver J, Miller JH. Regeneration beyond the glial scar. Nat Rev Neurosci. 2004;5(2):146–56.
36. Fawcett JW, Asher RA. The glial scar and central nervous system repair. Brain Res Bull. 1999;49(6):377–91.
37. Rolls A, Shechter R, Schwartz MNEURON-GLIAINTERACTIONS-OPINION. The bright side of the glial scar in CNS repair. Nat Rev Neurosci. 2009;10(3):235–U91.
38. Eddleston M, Mucke L. Molecular profile of reactive astrocytes--implications for their role in neurologic disease. Neuroscience. 1993;54(1):15–36.
39. Barreto GE, Morales L. Role of astrocytes in neurodegenerative diseases. https://doi.org/10.5772/31166
40. Yamanaka K, Chun SJ, Boillee S, Fujimori-Tonou N, Yamashita H, Gutmann DH, et al. Astrocytes as determinants of disease progression in inherited amyotrophic lateral sclerosis. Nat Neurosci. 2008;11(3):251–3.
41. Chu T, Zhou H, Li F, Wang T, Lu L, Feng S. Astrocyte transplantation for spinal cord injury: current status and perspective. Brain Res Bull. 2014;107:18–30.
42. Fields RD, Burnstock G. Purinergic signalling in neuron-glia interactions. Nat Rev Neurosci. 2006;7(6):423–36.
43. Zimmermann H. Signalling via ATP in the nervous system. Trends Neurosci. 1994;17(10):420–6.

44. Allan SM, Rothwell NJ. Cytokines and acute neurodegeneration. Nat Rev Neurosci. 2001;2(10):734–44.
45. Koller H, Thiem K, Siebler M. Tumour necrosis factor-alpha increases intracellular Ca2+ and induces a depolarization in cultured astroglial cells. Brain. 1996;119(Pt 6):2021–7.
46. Blanc EM, Bruce-Keller AJ, Mattson MP. Astrocytic gap junctional communication decreases neuronal vulnerability to oxidative stress-induced disruption of Ca2+ homeostasis and cell death. J Neurochem. 1998;70(3):958–70.
47. Hindley S, Herman MA, Rathbone MP. Stimulation of reactive astrogliosis in vivo by extracellular adenosine diphosphate or an adenosine A2 receptor agonist. J Neurosci Res. 1994;38(4):399–406.
48. Illes P, Norenberg W, Gebicke-Haerter PJ. Molecular mechanisms of microglial activation. B. Voltage- and purinoceptor-operated channels in microglia. Neurochem Int. 1996;29(1):13–24.
49. Abbracchio MP, Ceruti S, Bolego C, Puglisi L, Burnstock G, Cattabeni F. Trophic roles of P2 purinoceptors in central nervous system astroglial cells. Ciba Found Symp. 1996;198:142–7. discussion 7-8
50. Davalos D, Grutzendler J, Yang G, Kim JV, Zuo Y, Jung S, et al. ATP mediates rapid microglial response to local brain injury in vivo. Nat Neurosci. 2005;8(6):752–8.
51. Ajtai BM, Kalman M. Reactive glia support and guide axon growth in the rat thalamus during the first postnatal week. A sharply timed transition from permissive to non-permissive stage. Int J Dev Neurosci. 2001;19(6):589–97.
52. Kang W, Hebert JM. Signaling pathways in reactive astrocytes, a genetic perspective. Mol Neurobiol. 2011;43(3):147–54.
53. Ridet JL, Malhotra SK, Privat A, Gage FH. Reactive astrocytes: cellular and molecular cues to biological function. Trends Neurosci. 1997;20(12):570–7.
54. Faulkner JR, Herrmann JE, Woo MJ, Tansey KE, Doan NB, Sofroniew MV. Reactive astrocytes protect tissue and preserve function after spinal cord injury. J Neurosci. 2004;24(9):2143–55.
55. Fitch MT, Silver J. CNS injury, glial scars, and inflammation: inhibitory extracellular matrices and regeneration failure. Exp Neurol. 2008;209(2):294–301.
56. Mathewson AJ, Berry M. Observations on the astrocyte response to a cerebral stab wound in adult rats. Brain Res. 1985;327(1–2):61–9.
57. Koistinaho M, Lin S, Wu X, Esterman M, Koger D, Hanson J, et al. Apolipoprotein E promotes astrocyte colocalization and degradation of deposited amyloid-beta peptides. Nat Med. 2004;10(7):719–26.
58. Ikeda O, Murakami M, Ino H, Yamazaki M, Nemoto T, Koda M, et al. Acute up-regulation of brain-derived neurotrophic factor expression resulting from experimentally induced injury in the rat spinal cord. Acta Neuropathol. 2001;102(3):239–45.
59. Costa S, Planchenault T, Charriere-Bertrand C, Mouchel Y, Fages C, Juliano S, et al. Astroglial permissivity for neuritic outgrowth in neuron-astrocyte cocultures depends on regulation of laminin bioavailability. Glia. 2002;37(2):105–13.
60. Bush TG, Puvanachandra N, Horner CH, Polito A, Ostenfeld T, Svendsen CN, et al. Leukocyte infiltration, neuronal degeneration, and neurite outgrowth after ablation of scar-forming, reactive astrocytes in adult transgenic mice. Neuron. 1999;23(2):297–308.
61. Rothstein JD, Dykes-Hoberg M, Pardo CA, Bristol LA, Jin L, Kuncl RW, et al. Knockout of glutamate transporters reveals a major role for astroglial transport in excitotoxicity and clearance of glutamate. Neuron. 1996;16(3):675–86.
62. Swanson RA, Ying W, Kauppinen TM. Astrocyte influences on ischemic neuronal death. Curr Mol Med. 2004;4(2):193–205.
63. Chen Y, Swanson RA. Astrocytes and brain injury. J Cereb Blood Flow Metab. 2003;23(2):137–49.
64. Meyer JC, Geim AK, Katsnelson MI, Novoselov KS, Booth TJ, Roth S. The structure of suspended graphene sheets. Nature. 2007;446(7131):60–3.
65. Zador Z, Stiver S, Wang V, Manley GT. Role of aquaporin-4 in cerebral edema and stroke. Handb Exp Pharmacol. 2009;190:159–70.

References

66. McCarthy G, Leblond C. Radioautographic evidence for slow astrocyte turnover and modest oligodendrocyte production in the corpus callosum of adult mice infused with 3H-thymidine. J Comp Neurol. 1988;271(4):589–603.
67. Norton WT, Farooq M. Astrocytes cultured from mature brain derive from glial precursor cells. J Neurosci. 1989;9(3):769–75.
68. Miller R, Abney E, David S, Ffrench-Constant C, Lindsay R, Patel R, et al. Is reactive gliosis a property of a distinct subpopulation of astrocytes? J Neurosci. 1986;6(1):22–9.
69. Filous AR, Miller JH, Coulson-Thomas YM, Horn KP, Alilain WJ, Silver J. Immature astrocytes promote CNS axonal regeneration when combined with chondroitinase ABC. Dev Neurobiol. 2010;70(12):826–41.
70. Chen YSR. Astrocytes and brain injury. J Cereb Blood Flow Metab. 2003;23(2):137–412.
71. Harris FM, Tesseur I, Brecht WJ, Xu Q, Mullendorff K, Chang S, et al. Astroglial regulation of apolipoprotein E expression in neuronal cells implications for Alzheimer's disease. J Biol Chem. 2004;279(5):3862–8.
72. Verkhratsky A, Olabarria M, Noristani HN, Yeh CY, Rodriguez JJ. Astrocytes in Alzheimer's disease. Neurotherapeutics. 2010;7(4):399–412.
73. Cheng K, Kisaalita WS. Exploring cellular adhesion and differentiation in a micro–/nano-hybrid polymer scaffold. Biotechnol Prog. 2010;26(3):838–46.
74. Kliot M, Smith GM, Siegal JD, Silver J. Astrocyte-polymer implants promote regeneration of dorsal root fibers into the adult mammalian spinal cord. Exp Neurol. 1990;109(1):57–69.
75. Wang JJ, Chuah MI, Yew DT, Leung PC, Tsang DS. Effects of astrocyte implantation into the hemisected adult rat spinal cord. Neuroscience. 1995;65(4):973–81.
76. Joosten EA, Veldhuis WB, Hamers FP. Collagen containing neonatal astrocytes stimulates regrowth of injured fibers and promotes modest locomotor recovery after spinal cord injury. J Neurosci Res. 2004;77(1):127–42.
77. Davies JE, Huang C, Proschel C, Noble M, Mayer-Proschel M, Davies SJ. Astrocytes derived from glial-restricted precursors promote spinal cord repair. J Biol. 2006;5(3):7.
78. Davies JE, Proschel C, Zhang N, Noble M, Mayer-Proschel M, Davies SJ. Transplanted astrocytes derived from BMP- or CNTF-treated glial-restricted precursors have opposite effects on recovery and allodynia after spinal cord injury. J Biol. 2008;7(7):24.
79. Liu Y, Wang L, Long Z, Zeng L, Wu Y. Protoplasmic astrocytes enhance the ability of neural stem cells to differentiate into neurons in vitro. PLoS One. 2012;7(5):e38243.
80. Luo L, Guo K, Fan W, Lu Y, Chen L, Wang Y, et al. Niche astrocytes promote the survival, proliferation and neuronal differentiation of co-transplanted neural stem cells following ischemic stroke in rats. Exp Ther Med. 2017;13(2):645–50.
81. Bacigaluppi M, Russo GL, Peruzzotti-Jametti L, Rossi S, Sandrone S, Butti E, et al. Neural stem cell transplantation induces stroke recovery by upregulating glutamate transporter GLT-1 in astrocytes. J Neurosci. 2016;36(41):10529–44.
82. Becerra-Calixto A, Cardona-Gómez GP. Neuroprotection induced by transplanted CDK5 knockdown astrocytes in global cerebral ischemic rats. Mol Neurobiol. 2016:1–16.
83. Cao Q, Xu X-M, DeVries WH, Enzmann GU, Ping P, Tsoulfas P, et al. Functional recovery in traumatic spinal cord injury after transplantation of multineurotrophin-expressing glial-restricted precursor cells. J Neurosci. 2005;25(30):6947–57.
84. Xu L, Yan J, Chen D, Welsh AM, Hazel T, Johe K, et al. Human neural stem cell grafts ameliorate motor neuron disease in SOD-1 transgenic rats. Transplantation. 2006;82(7):865–75.
85. Lepore AC, Rauck B, Dejea C, Pardo AC, Rao MS, Rothstein JD, et al. Focal transplantation–based astrocyte replacement is neuroprotective in a model of motor neuron disease. Nat Neurosci. 2008;11(11):1294–301.
86. Lepore AC, O'Donnell J, Kim AS, Williams T, Tuteja A, Rao MS, et al. Human glial-restricted progenitor transplantation into cervical spinal cord of the SOD1G93A mouse model of ALS. PLoS One. 2011;6(10):e25968.
87. Li K, Javed E, Scura D, Hala TJ, Seetharam S, Falnikar A, et al. Human iPS cell-derived astrocyte transplants preserve respiratory function after spinal cord injury. Exp Neurol. 2015;271:479–92.

88. Nicaise C, Mitrecic D, Falnikar A, Lepore AC. Transplantation of stem cell-derived astrocytes for the treatment of amyotrophic lateral sclerosis and spinal cord injury. World J Stem Cells. 2015;7(2):380.
89. Potts MB, Silvestrini MT, Lim DA. Devices for cell transplantation into the central nervous system: design considerations and emerging technologies. Surg Neurol Int. 2013;4(Suppl 1):S22.
90. Habisch H-J, Janowski M, Binder D, Kuzma-Kozakiewicz M, Widmann A, Habich A, et al. Intrathecal application of neuroectodermally converted stem cells into a mouse model of ALS: limited intraparenchymal migration and survival narrows therapeutic effects. J Neural Transm. 2007;114(11):1395–406.
91. Kim H, Kim HY, Choi MR, Hwang S, Nam K-H, Kim H-C, et al. Dose-dependent efficacy of ALS-human mesenchymal stem cells transplantation into cisterna magna in SOD1-G93A ALS mice. Neurosci Lett. 2010;468(3):190–4.
92. Corti S, Locatelli F, Donadoni C, Guglieri M, Papadimitriou D, Strazzer S, et al. Wild-type bone marrow cells ameliorate the phenotype of SOD1-G93A ALS mice and contribute to CNS, heart and skeletal muscle tissues. Brain. 2004;127(11):2518–32.
93. Ohnishi S, Ito H, Suzuki Y, Adachi Y, Wate R, Zhang J, et al. Intra-bone marrow-bone marrow transplantation slows disease progression and prolongs survival in G93A mutant SOD1 transgenic mice, an animal model mouse for amyotrophic lateral sclerosis. Brain Res. 2009;1296:216–24.
94. Berndt M, Li Y, Seyedhassantehrani N, Yao L. Fabrication and characterization of microspheres encapsulating astrocytes for neural regeneration. ACS Biomater Sci Eng. 2017;3(7):1313–21.
95. Takahashi H, Itoga K, Shimizu T, Yamato M, Okano T. Human neural tissue construct fabrication based on scaffold-free tissue engineering. Adv Healthc Mater. 2016;5(15):1931–8.
96. East E, de Oliveira DB, Golding JP, Phillips JB. Alignment of astrocytes increases neuronal growth in three-dimensional collagen gels and is maintained following plastic compression to form a spinal cord repair conduit. Tissue Eng Part A. 2010;16(10):3173–84.
97. Liddelow SA, Guttenplan KA, Clarke LE, Bennett FC, Bohlen CJ, Schirmer L, et al. Neurotoxic reactive astrocytes are induced by activated microglia. Nature. 2017;541(7638):481–7.
98. Zhang Y, Sloan SA, Clarke LE, Caneda C, Plaza CA, Blumenthal PD, et al. Purification and characterization of progenitor and mature human astrocytes reveals transcriptional and functional differences with mouse. Neuron. 2016;89(1):37–53.
99. Glover JC, Boulland J-L, Halasi G, Kasumacic N. Chimeric animal models in human stem cell biology. ILAR J. 2010;51(1):62–73.
100. Lindvall O, Kokaia Z. Stem cells in human neurodegenerative disorders—time for clinical translation? J Clin Invest. 2010;120(1):29.
101. Allard J, Li K, Lopez XM, Blanchard S, Barbot P, Rorive S, et al. Immunohistochemical toolkit for tracking and quantifying xenotransplanted human stem cells. Regen Med. 2014;9(4):437–52.

Chapter 2
Enhancement of Axonal Myelination in Wounded Spinal Cord Using Oligodendrocyte Precursor Cell Transplantation

Li Yao and Michael Skrebes

Abstract Oligodendrocytes (OLs) myelinate axons in the central nervous system (CNS). OLs form myelin sheath which is a spiral structure around a neuron axon. Each OL forms multiple segments of myelin that wrap around many spinal axons. The myelination increases the propagation rate of action potentials through the axon, thereby allowing more efficient signal transmission through the nervous system. Spinal cord injury causes massive tissue destruction and vascular rupture. Such trauma causes the loss of oligodendrocytes and demyelination. Following CNS injury, OLs proliferate and migrate into the neural lesion and limited myelin regeneration. A number of genes that are activated and involved in process of remyelinating damaged axons have been identified. The transplantation of OPCs has been shown as an effective treatment in restoring remyelinated neurons. A biomaterial scaffold can act as a vehicle for OPC delivery and create a permissive environment for transplanted cells to improve axonal myelination. It was also demonstrated that electric fields (EFs) can direct OPCs migration. This emerging treatment may be used as a novel therapeutic method to enhance OPCs migration toward the lesion.

Keywords Oligodendrocyte precursor · Central nervous system · Axon · Migration · Neural regeneration · Spinal cord injury · Biomaterials scaffolds · Nanofiber · Microenvironment · Transplantation

2.1 Axonal Myelination in Central Nervous System

In the central nervous system (CNS), axons are myelinated by neuroglial oligodendrocytes (OLs) cells. The myelin sheath formed by OLs is a spiral structure around a neuron axon. Each OL forms multiple segments of myelin that wrap around many spinal axons. The primary function of myelination is to increase the propagation rate of action potentials through the axon, thereby allowing more efficient signal transmission through the nervous system. The myelin sheath insulates the axon from electrical activity, and the segmented structure of the sheath allows for rapid

© The Author(s) 2018
L. Yao, *Glial Cell Engineering in Neural Regeneration*,
https://doi.org/10.1007/978-3-030-02104-7_2

saltatory conduction of nerve impulses across segments. Neuronal myelination is essential to the maintenance of proper function within the human brain and spinal cord. Deficiencies in myelination can inhibit muscle control, cognitive function, and social development. Proper myelination is essential for neuronal connectivity and higher cognitive function within the brain.

Myelination in healthy individuals is an ongoing process beginning early in prenatal development and continuing well into the later stages of life [1]. The developmental path of myelination begins with oligodendrocyte precursor cells (OPCs), which form in the ventricular zone of the embryonic nervous system and then migrate through the CNS [2]. OPCs migrate long distances to participate in the formation of spinal cord white matter [3, 4]. During embryonic development, OPCs distribute evenly by migrating along the vascular endothelium [2]. In the process of migration, they attach to vascular elements, and a leading process guides them along the vascular elements. After OPCs reach their target cells, they can then mature into OLs and begin myelinating neuron axons [2].

When oligodendrocytes mature, they extend a covering, known as the myelin sheath, across the outer axonal membrane. The mature OL wraps the axon concentrically as a multilamellar sheet of plasma membrane. After attachment, the myelin sheath continues to grow along with the development of the axon, a process facilitated by cytoskeletal rearrangement within the OL [5]. The myelination of axons speeds up the propagation of action potentials through the neurons of the CNS [6, 7].

The migration of OPCs, their maturation into OLs, and the final process of myelination are all dependent on intrinsic mechanisms and extracellular regulators. Although poorly understood, chemical regulators in the extracellular matrix have been shown to be important control factors in support or inhibition of the myelination process. Growth factors such as the fibroblast growth factor (FGF) and platelet-derived growth factor (PDGF) inhibit OPC maturation, while conversely, their removal allows for maturation [8, 9].

2.2 Axon Demyelination Resulting from Spinal Cord Injury

The myelination process is dynamic and ongoing throughout most of life. Oligodendrocyte progenitors maintain the capability of continuous generation in the CNS. Demyelination is a process whereby myelin is lost while the axon remains largely intact. In cases of demyelination occurring in a healthy individual, the OLs can restore the myelin sheath to equilibrium. In the adult CNS, OPCs retain the potential to regenerate and the capacity for myelination after spinal cord injury (SCI) [10]. In cases resulting in the loss of OLs, such as could occur in SCI, or a disruption in maturation, as could occur in degenerative disease such as multiple sclerosis and progressive multifocal leukoencephalopathy, the reduction of neurological conductivity has been marked as an important regulator of symptomatic conditions [6, 11]. Mechanical injury to the spinal cord can cause massive tissue destruction and vascular rupture. Such trauma causes acute death of neurons and oligodendrocytes. Demyelination occurs after the OLs undergo both necrosis and apoptosis at the site of injury. Pathophysiological changes in the secondary injury lead to continuous OL loss and demyelination. The cascade reaction at the cellular and molecular level in the second injury may last several days or weeks [12–14]. Oligodendrocyte cell loss was detected at an early stage of SCI in animal models. After an incomplete midthoracic contusion injury of a rat spinal cord, the loss of OLs in the ventral white matter was observed at the lesion 4 h after SCI [15]. Significant OL apoptosis in degenerating axon tracts distant to the lesion lasted for at least 3 weeks after a contusion injury in a rat animal model [16, 17]. The demyelination is accompanied by Wallerian degeneration, which can occur in the axons distal to the lesion of injury. Oligodendrocytes will lose tropic support after axonal degeneration. The activated microglia are also involved in OL apoptosis that cause demyelination of axons [18].

The chemical and molecular environment in the lesion causes oligodendrocyte loss and demyelination. The over-releasing of glutamate, which is an excitatory neurotransmitter, is toxic to OLs post-SCI. Nitrosative and oxidative stresses also can trigger mitochondrial fragmentation and induce cell death [19, 20]. Nitric oxide (NO) is generated by the vascular endothelium and neurons in the CNS. A small amount of free radicals, such as superoxide and hydroxyl, are produced by the mitochondrial electron transport chain and the microsomal cytochrome P-450 system. Ischemia and reperfusion also enhance the generation of free radicals in the wounded spinal cord. Caspases are the main cysteine proteases for execution of apoptosis, and they participate in the process of oligodendrocyte apoptosis in SCI. Calpains are Ca^{2+}-dependent cysteine proteases that can cause both necrotic and apoptotic death in SCI [21, 22]. The inhibition of the function of these molecules can prevent cell apoptosis and OL loss. The function of caspases 3, 8, and 9 is primarily activated in neurons and OLs, and intrathecal injection of the caspase inhibitor improves rat locomotor function [23]. Administration of the calpain-specific inhibitor to injured rat spinal cords continuously for 24 h inhibited calpain activity and other factors that lead to apoptosis in the lesion and surrounding areas. Outcomes of the study obtained from the SCI animal model suggest that the calpain inhibitor can provide neuroprotection for human patients with SCI [21].

2.3 Restoration of Axonal Myelination Post-Spinal Cord Injury

The response of glial cells following injury of the CNS has been observed to have both regenerative and destructive functionality. The response to CNS damage initially begins with the migration of microglia to the damaged area [24]. After microglia reach the lesion, they undertake a process of cleanup, quarantining apoptotic cells and cellular debris. Following the initial cleanup, astrocytes begin to accumulate and eventually form a layer of scar tissue around the damaged axon [25]. The response of glial cells to CNS injury has shown that they improve the functionality of damaged neurons while at the same time playing a role in limiting the overall degree of CNS recovery. The scar tissue layer formed by astrocytes around the damaged neuron serves first to quarantine the damaged area but then inhibits regrowth in the damaged axon [24]. Overactivation of microglia responding to the injury site poses a threat to OPCs and mature OLs [26]. Responding microglia have been observed to play an important role in removing cellular components that can prevent myelin restoration, but they have also been shown to nonspecifically target healthy OPCs and OLs, thus restricting the restoration of myelin function. The function of nerve/glial antigen 2 (NG2)-expressing progenitors changes according to alternations in the microenvironment post-SCI. The NG2-expressing progenitors have shown a phenotypic shift between an astroglial-generating glial scar and the myelinating OLs [27]. One study showed that NG2 progenitors born 24 h after SCI mainly generated scar-forming astrocytes, while NG2 progenitors born 7 days after injury generated mature, myelinating OLs [27].

Limited myelin regeneration has been observed following CNS injury [28–30]. The OPCs presenting throughout adult gray and white matter of spinal cord are activated in response to the demyelination post-SCI and proliferate and migrate into the neural lesion [31]. The subventricular zone (SVZ) is another region in the adult CNS that generates OPCs, which can then migrate to the lesion and participate in the myelination process [28, 29]. However, most proliferating progenitors are astrocytes, and few cells are oligodendrocytes [32, 33]. Adult OPCs typically express NG2 or platelet-derived growth factor receptor (PDGFRα). These findings were confirmed by a spinal cord contusion study [33]. Although cell proliferation was observed from 1 day post-SCI, very few of these cells expressed NG2. The cellular response to CNS injury contrasts with that of peripheral nervous system (PNS) injury in that it is unable to fully regenerate the damaged area and instead works only to partially restore myelin function.

Schwann cells (SCs) myelinate nerve in the PNS by regenerating and migrating to the injury site post-SCI to wrap the spinal cord axons [34, 35]. A study of peripheral and central myelin at the chronical spinal cord lesion of postmortem human patients revealed that SCs contributed to the restoration of myelin sheaths around some spinal axons [35]. Some SCs derived from progenitor cells in the CNS can form the peripheral-like myelin [36–38]. When neural precursor cells derived from adult human brain tissue removed during surgery were cultured, a small number of cells showed the SC phenotype. With the transplantation of neural precursor cells into the demyelinated adult rat spinal cord, extensive remyelination can occur, demonstrating a peripheral myelin pattern similar to Schwann cell-formed myelination.

The basic mechanism of axonal myelination in early development of the central nervous system has been discerned, but the intricacies of its continuation through adulthood are still poorly understood. OPCs emerging from progenitor domains of the embryonic mouse brain and spinal cord and the human cortex are associated with the abluminal endothelial surface of nearby blood vessels. The cells migrated within the vascular network. The role of the Wnt-CXCR4 signaling pathway in OPC migration toward target cells has marked it as an important mechanism in enabling the cellular targeting of myelination to specific cells [39, 40]. Recent research has dem-

2.3 Restoration of Axonal Myelination Post-Spinal Cord Injury

onstrated that CXCR4 was expressed by OPCs during embryonic developmental migration but was down-regulated along with the Wnt pathway down-regulation in differentiating mature OLs. Wnt activation in OPCs mediated their attraction to the vasculature during migration. The Wnt down-regulation is required for appropriate endothelial dissociation and the subsequent differentiation of OPCs into OLs [40].

Differentiation of OPCs is regulated by growth factors, substrate signaling, and chemokines [3, 10, 41, 42]. Growth factors such as PDGF and basic FGF (bFGF) can improve cell proliferation by promoting OPC division, while transforming growth factor beta (TGF-β) inhibits the proliferation of OPCs and induces cell maturation. The proliferative response of spinal cord OPCs to PDGF depends on the chemokine CXCL1 [43, 44]. CXCL1 inhibited the migration of OPCs and widespread dispersal of post-natal precursors observed in slices of CXCR2 knockout animals [41]. Notch signaling regulates the differentiation of OLs spatially and temporally in the CNS [45]. In an in vitro study, it was shown that Notch signaling can inhibit the OPC differentiation [46].

The continued process of myelination beyond prenatal development is important to understanding SCI recovery. While some function can be restored in the CNS, the regenerative capacity of OLs in their natural state is limited. In response to neural injury, OPCs can be activated and can initiate the myelination process. However, underlying mechanisms are the focus of much research. Several genes have been identified as playing a role in the process of remyelinating damaged axons including Notch1, NG2, prox1, TNF, and NFκB [25]. The signaling interaction between these genes leads to a surge in the NG2 proliferation in response to injury and ensuing differentiation. The deregulation of the G-protein coupled receptor (GPCR), such as protease precursors of protease activated receptor 2 (PAR2), is a common feature expressed after neurological injury and disease [11, 47]. The deletion of PAR2 has been linked to increases in Olig-2 and CC1-positive oligodendrocytes [11]. The increase in myelination efficiency has been confirmed through experiments utilizing a PAR2 knockout mice model. The myelin regeneration rate was increased when a PAR2 knockout mouse was subjected to traumatic spinal injury and subsequent demyelination, in comparison to a control mouse. These studies shed important light on the potential mechanisms that allow for the propagation of oligodendrocytes and therefore contribute to the development of novel therapeutic strategies. Chromodomain helicase DNA binding protein 7 (CHD7) is also involved in the regulation of OPC activation after SCI [48–50]. The reduced OPC proliferation, loss of OPC identity, and impaired OPC differentiation were observed in the mice of SCI with OPC-specific knockout of CHD7. The function of CHD7 in regulating OPC activation acts via direct induction of the regulator of cell cycle (RGCC) and protein kinase Cθ (PKCθ) expression.

2.4 Transplantation of OPCs Enhancing Axon Remyelination in Spinal Cord Injury

2.4.1 Transplantation of OPCs in Therapy of SCI

In response to demyelination, although gray and white matter maintains the capacity to generate oligodendrocyte precursor cells that can migrate to the lesion to remyelinate axons, the limited availability of these endogenous OPCs is insufficient for remyelination in the adult spinal cord [31, 51]. The transplantation of OPCs has been shown as an effective treatment in restoring remyelinated neurons back toward natural equilibrium. Studies have shown that OPCs can survive, proliferate, and migrate after transplantation into the host neural tissue [52]. It was also reported that the survival of transplanted OPCs can be supported further by introducing the tumor necrosis factor-alpha antagonist [53]. After the OPCs isolated from the fetal human forebrain and adult white matter were grafted into the forebrains of newborn shiverer mice, both fetal and adult OPCs mediated the extensive and robust myelination of congenitally dysmyelinated mice brain [54]. After tissue fragments isolated from the brains and spinal cords of 10-week-old human fetuses were grafted into the brains of newborn shiverer mutant mice, the OPCs of the grafted tissue migrated into the host neural tissue and myelinated the axons [55]. The transplantation of OPCs can also enhance remyelination and functional recovery after SCI. The transplantation of OPCs into rat spinal cord suffering contusive injury [52] significantly improved the locomotor function and motor-evoked potential (MEP) of the wounded animal. In one study, OPCs were infected with retroviruses expressing ciliary neurotrophic factor (CNTF) and then transplanted into the contused rat thoracic spinal cord. The grafted OPCs survived and were integrated into the injured spinal cord. The grafted OPCs formed central myelin sheaths around the axons and significantly enhanced the recovery of hind limb locomotor function. Glial-restricted precursor cells (GRPs) isolated from an embryonic day-14 rat spinal cord can be differentiated into both oligodendrocytes and astrocytes [56]. After transplantation of GRPs infected with retroviruses expressing brain-derived neurotrophic factor (BDNF) and neurotrophin-3 (NT-3) into contused thoracic spinal cord, the grafted GRPs differentiated into mature OLs and formed myelin sheaths around the axons. The locomotor function recovery of the hind limb was significantly improved [56].

A recent study revealed that OPC transplantation promoted functional recovery of rats with contusive SCI, and the improved functional recovery was associated with the altered expression of various miRNAs in the spinal cord [57]. The most highly up-regulated miRNA (miR-375-3p and miR-1-3p) and down-regulated miRNA (miR-36-3p, miR-449a-5p, and miR-3074) were identified. The molecules MiR-375 and miR-1 may inhibit cell proliferation and apoptosis, while miR-363, miR-449a, and miR-3074 inhibit cell proliferation and neuronal differentiation.

The transplantation of oligodendrocyte precursor cells has been shown to enhance remyelination of axons in the CNS lesion in in vivo studies. However, the clinical application of human OPCs in spinal cord repair has been limited by the available OPC source and the low harvest of cultured OPCs. Oligodendrocyte precursor cells derived from stem cell differentiation have provided an alternative source of cells for transplantation in SCI [58, 59]. Those derived from embryonic stem cells (ESCs) and neural stem cells (NSCs) have been investigated for focal delivery to remyelinate axons [60–62]. In one study, transplanted human embryonic stem cell (hESC)-derived OPCs in a wounded spinal cord survived and migrated over short distances in the lesion [62]. In another study, the transplantation of hESC-derived OPCs into a completely transected spinal cord improved locomotor function based on the evaluation of Basso–Beattie–Bresnahan locomotor scores [61].

Olfactory unsheathing cells (OECs) provided an alternative source of cells to treat SCI. To improve myelination and restore the functional recovery of the injured spinal cord, they were transplanted into a patient in clinical trials. Transplantation of OEC cells into SCI has been conducted in a few clinical studies [63–66]. In a study by Mackay-Sim et al. [63], OECs isolated from the nasal mucosa of six patients with chronic SCI were transplanted back into the injured spinal cords. Although no significant functional changes were reported by these patients, the cell transplantation did not show adverse effects such as neuropathic pain 3 years after transplantation. In another study, olfactory mucosa autografts (OMAs) were transplanted into 20 patients with chronic, sensorimotor complete, or motor complete SCI after a partial scar was removed [66]. The ASIA Impairment Scale (AIS) improved in 11 of 20 patients, and these improvements included new voluntary electromyography (EMG) responses (15 patients) and somatosensory-evoked potentials (SSEPs) (4 patients). This study indicated that OMA transplantation is a feasible and relatively safe procedure and that it may be beneficial in people with chronic SCI when combined with post-operative rehabilitation.

2.4.2 Co-Transplantation of Biomaterial Scaffolds and OPCs for SCI Therapy

The microenvironment of transplanted oligodendrocyte precursor cells in the spinal cord lesion is a hostile environment for cell survival and differentiation. Implantable biomaterial-encapsulating cells can generate a permissive environment for the survival of grafted cells. Studies have shown that the wounded spinal cord reconstructed by bridging the gap with neural conduits or filling the defect with hydrogels exhibited improved neural regeneration [67–69]. A biomaterial scaffold that conducts the neural tissue growth can also act as a vehicle for OPC delivery to enhance axonal myelination [70]. To generate the optimal condition, biomaterial scaffolds should have appropriate mechanical durability, pore size, capsule consistency, degradation features, and low immunogenicity [70]. Implantation of biomaterial scaffolds seeded with stem cells such as NSCs, mesenchymal stem cells (MSCs), and hESCs can improve axonal regeneration, reduce glial scar formation, and promote functional recovery post-SCI. It was reported that an injectable chitosan sponge cross-linked with guanosine 5′-diphosphate (GDP) was fabricated to investigate

OPC survival, attachment, and differentiation [71]. OPCs can be differentiated into mature oligodendrocytes that express myelin basic protein (MBP) when cultured on sponges containing NT-3.

Micro- and nanoscale particles synthesized from various natural and synthetic polymers to encapsulate biomolecules or therapeutic cells have been investigated. Biomolecules encapsulated in particles may generate a significant biological effect through a small volume [49, 72–74]. Biodegradable microspheres can act as efficient tools for the controlled release of these molecules. The sustained release of vascular endothelial growth factor (VEGF) from collagen microspheres was reported [75]. This growth factor, released from collagen microspheres over the course of 4 weeks, retained its bioactive properties. When fibrin containing poly(lactic-c-glycolic acid) microsphere-encapsulated glial-derived neurotrophic factor (GDNF) was introduced into a wounded rat nerve, the released GDNF enhanced nerve regeneration [76]. Cell delivery by microspheres and transplantation to the targeted tissue may be an efficient approach for cell therapy. Previous studies have shown that microspheres are suitable for stem cell growth and can potentially deliver stem cells to the injured area for tissue regeneration [77–79]. Microspheres may provide an optimal condition for the delivered cells to adapt to the hostile environment of the SCI lesion. Biodegradable and biocompatible natural polymers are strongly favored for the cell delivery. Because of their elastic property, the local delivery of collagen-based microspheres to neural tissue by injection causes little damage to healthy tissue. Adipose-derived stem cells (ADSCs) can grow on porous chitosan microspheres, and these cells proliferated on the surface of the spheres and infiltrated the pores to grow within the spheres [78]. When microencapsulated islets were implanted into an animal with diabetes, the microencapsulated cells remained viable and controlled glucose levels for several weeks in the hosts [80, 81]. In one study, collagen microspheres were fabricated and acted as a carrier of the OPCs. Myelination was observed in the co-culture of OPCs on microspheres with dorsal root ganglia (DRG) [82] (Fig. 2.1). Cells delivered by collagen-based microspheres may constantly generate therapeutic molecules into the injury site. In one study, astrocytes that were transfected with nerve growth factor (NGF)-enhanced green fluorescent protein (EGFP) plasmids were encapsulated in collagen microspheres. Most NGF that is generated by astrocytes was released from the collagen microspheres, and the NGF significantly promoted axonal growth of the cultured DRG [83].

Nanofibers can mimic the extracellular matrix and provide guidance for axonal growth at nanolevels. The growth of dorsal root ganglia neurons can be guided by electrospun aligned fibers in vitro [84–86]. Nanofibers grafted into an injured spinal cord and peripheral nerve enhanced axonal regeneration [87–90]. The growth of multiple glial cell populations including astrocytes, OPCs, and OLs can be guided by aligned nanofibers. The parallel nanofibers induced an oriented growth of astrocytes, and the growth, elongation, and maturation of OPCs required prealigned astrocytes. Although the co-culture of neurons and OPCs is the general model for myelination studies, the intimate interaction between axons and OL cells is a complicated process [91–93]. Nerve growth factor and serum in the cell culture medium negatively affect the myelination function of OL cells and increase the non-specific

2.4 Transplantation of OPCs Enhancing Axon Remyelination in Spinal Cord Injury

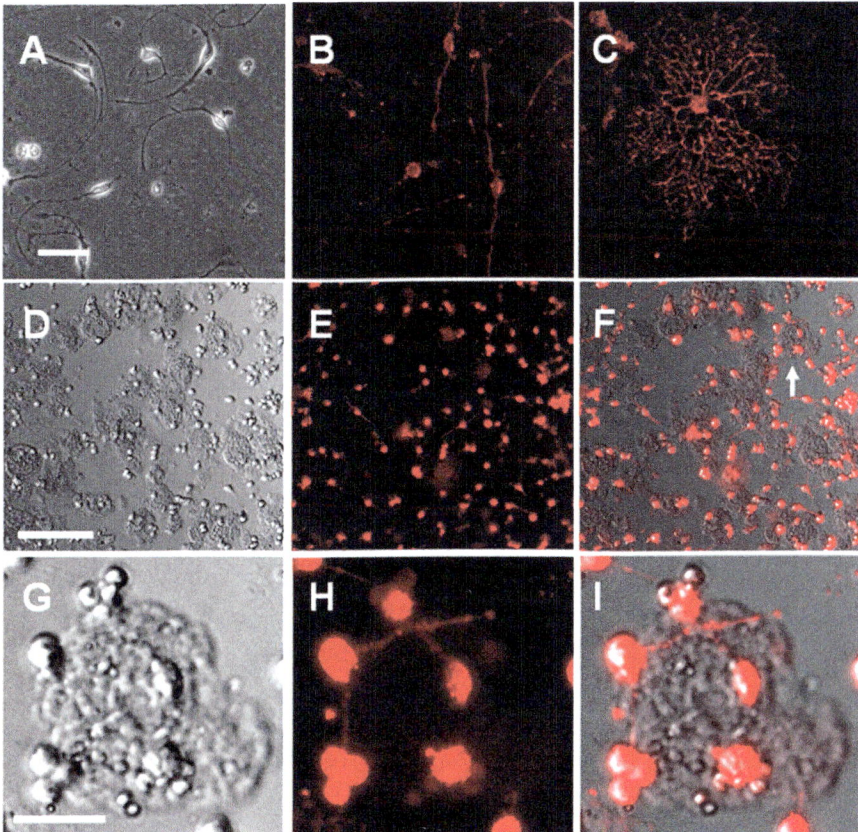

Fig. 2.1 Growth of OPCs on collagen microspheres: (**a–c**) Oligodendrocyte progenitor cells (OPCs) grown in cell culture plate labeled with anti-A2B5 antibody. Scale bar: 100 µm. (**d–f**) Bright field and fluorescent images showing OPCs grown on collagen microspheres and labeled with anti-A2B5 antibody. Scale bar: 200 µm. (**g–i**) Typical microsphere with growth of OPCs highlighting the morphology of OPCs grown on microspheres. Scale bar: 50 µm. Figure reproduced from Yao et al. [82] with permission from BioMed Central

cells. The nanofibers that mimic axon morphology and diameter can act as a neuron-free model to study myelination. Studies have shown that OPCs can myelinate electrospun fibers of synthetic polymers [94, 95]. When OPCs were grown on PCL-gelatin copolymer nanofibers, OLs formed significantly more myelinated segments than those on PCL nanofibers alone. This study suggested that the function of a particular biological molecule in the myelination process can be studied by incorporating it into fibers. The myelination of nanofibers is a neuron-free model that simplifies the process and focuses on the interaction of particular biological molecules and OPCs [96] (Fig. 2.2).

In another study, nanofibers were functionalized with microRNA (miR), which controlled oligodendrocyte precursor cell differentiation through gene silencing [97]. Nanofiber-mediated delivery of miR-219 and miR-338 promoted oligodendro-

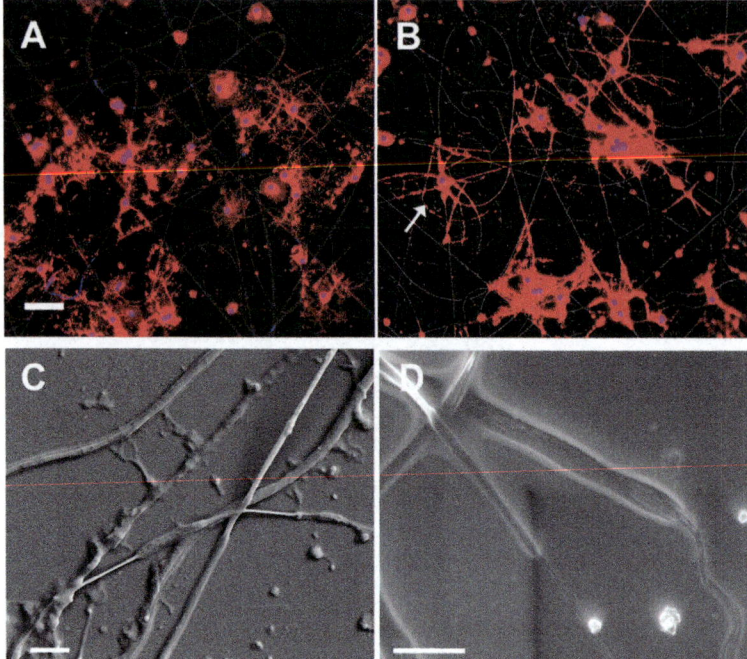

Fig. 2.2 Differentiated OPC-ensheathed nanofibers: (**a**) Differentiated OPC-ensheathed PCL fibers. (**b**) Differentiated OPC-ensheathed PCL-gelatin fibers. Scale bar: 100 μm. (**c**, **d**) SEM images showing differentiated OPC-ensheathed PCL-gelatin nanofibers. Figure reproduced from Li et al. [96] with permission from American Chemical Society

cyte maturation by significantly increasing the number of MBP cells. Results demonstrated that both topographical cues of nanofibers and microRNA reverse transfection can efficiently direct OPC differentiation. The scaffolds functionalized with biological molecules may be applied in directing oligodendrocyte differentiation and myelination for treatment of CNS pathological conditions.

2.5 Potential Application of Electrical Stimulation in Axonal Myelination

Neuronal myelination begins at the stage of postnatal development and continues throughout most of life. This myelination process contributes to the continued neural myelination and plasticity, and requires the continued maturation of oligodendrocyte precursor cells into oligodendrocytes [2]. Research of adaptive myelination has suggested that an electrical impulse stimulus is important to continued myelination [98]. Studies have shown that the restriction of social interaction of two-week-old puppies raised under isolation conditions leads to the formation of thinner and fewer myelin sheaths [99, 100]. The restricted progression of myelination in the puppy model was corrected 4 weeks after the puppies were removed from isolation, demonstrating the importance of stimulus reception to brain development and neural plasticity.

Research exploring the use of electrical stimulation as a means of neuronal stimulation has shown promise in stimulating neural plasticity and in possibly reversing the process of demyelination. In one in vitro study, mice dorsal root ganglion neurons and OPCs were co-cultured on a microfluidic chamber and subjected to electrical stimulation. Results showed that electrical stimulation of neurons enhances oligodendrocyte maturation and myelin formation. The enhanced myelination was independent of the input localization and oligodendrocyte exposure to the electrical field [98]. In a further in vivo study, epidural electrodes were implanted over the primary motor (M1) cortex in the contusive SCI of rodent model at T10 level and the neuronal activity was induced by electrical stimulation. The induced neuronal activity also increased the number of proliferating OPCs, the number of oligodendrocytes, and myelin sheath formation in the dorsal corticospinal tract (dCST) in the subchronic stages of SCI. This study suggested that epidural electrical stimulation that induces neuronal activity throughout white matter tracts of the CNS could be used to promote remyelination after CNS injury and demyelination disorders [101].

A number of previous studies have demonstrated that an electrical field can guide the migration of primary neurons and neural stem cells. In one study, the migration of OPCs was directed to the anode pole, whereby the increased voltage enhanced the directedness and net displacement of anodal migration of the OPCs. Similar to OPCs, Schwann cells migrated to the anode when stimulated with an electrical field (EF) [102, 103]. Rat hippocampal neurons showing bipolar morphology at an early culture stage are very motile. In contrast to OPCs and Schwann cells, their migration can be guided toward the cathode pole in EFs. Motor neuron-derived embryonic stem cells also showed bipolar morphology with short processes, and they migrated to the cathode in EFs [104]. These studies indicate that EFs may function as a guidance cue to direct the migration of endogenous or transplanted primary neural cells to the lesion to establish the physiological connection.

Although an electrical field directs neural cell migration, it does not affect the motility of most cell types. The migration speeds of hippocampal neurons, embryonic stem cell-derived motor neurons, Schwann cells, and OPCs did not significantly change in EFs. The cell migration velocity of these cells was about 60 µm/h. However, it was observed that the migration speed of hiPSCs was increased by EF stimulation [105]. These studies suggest that cell motility and migratory direction are regulated differently by EF stimulation. Thresholds of EF stimulation for guided cell migration varied among cell types. The EF strength that initiated directional migration of avian neural crest cells was 7 mV/mm [106, 107]. The threshold of EF-directed migration of hippocampal neurons and Schwann cells was about 100 mV/mm. In this study, we found clear anodal migration of OPCs in EFs of 200 mV/mm but insignificantly directed migration in EFs of 100 mV/mm. This indicates that the threshold effectiveness of OPCs should lie between these two values.

Though the electrical field can clearly direct neural cell migration, the mechanism regulating cell migration is still far from being fully understood. RNA sequencing (RNA-Seq) was used to systematically detect differentially expressed genes (DEGs) by comparing OPCs subjected to EF stimulation and control cells. DEGs using gene ontology (GO) terms and the Kyoto Encyclopedia of Genes and Genomes (KEGG) pathway analysis were studied. KEGG pathway analysis was used to explore the significantly changed molecular pathways, and a number of genes that may be important in the regulation of EF-guided cell migration was also explored.

The interaction of chemokine and chemokine receptors regulates the directional cell migration in chemotaxis. Chemokines are involved in spinal cord development by regulating the migration of OPCs. The interaction of chemokine CXCL1 and its receptor CXCR2 patterned development of the spinal cord. CXCL1 inhibited OPC migration by signaling its receptor CXCR2 [41]. The RNA-Seq study of OPCs revealed that DEGs were significantly enriched in taxis and chemotaxis. The chemokine genes—CCL3, CCL4, CXCL1, and CXCL2—were significantly down-regulated for cells subjected to EFs of 100 mV/mm compared with control cells. CXCl2 was down-regulated when the cells were subjected to a higher EF strength (200 mV/mm). Results indicate that the generation of chemokines may be suppressed by EFs, and chemotaxis and electrotaxis may be regulated differently.

The mitogen-activated protein kinase (MAPK) pathway regulates various cellular functions such as cell proliferation, differentiation, and migration. It was reported that this pathway mediated EF-directed cell migration. EF stimulation can activate the extracellular signal-regulated kinase (ERK) in corneal epithelial cells and cause the accumulation of F-actin at the leading, cathodal-facing side of the cell [108]. In the RNA-Seq study, a number of genes of the MAPK pathway including FOS, DUSP2, GADD45A, GADD45B, GADD45G, and NR4A1 were up-regulated in OPCs stimulated with an EF of 200 mV/mm and compared with control cells. Previous studies have shown that these molecules regulated cell migration. GADD45A regulated β-catenin subcellular distribution and cell adhesion [109]. One recent study [110] demonstrated that GADD45A null mouse embryonic fibroblast (MEF) cells showed significantly increased adhesion and migratory ability in vitro. NR4A1 regulates cell migration and acts as an antimigratory factor in normal mammary epithelial and breast cancer cell lines [111]. NR4A1-overexpression in normal epithelial cells and breast cancer cells reduced cell migration ability. PAI-1 is a serine proteinase inhibitor (serpin) and mediates cell migration by modulating interactions between vitronectin, and either the uPA receptor (uPAR) or integrins [112–115]. FOS proteins may regulate cell motility. Antisense C-FOS transcripts effectively reduced PAI-1 induction and growth factor-stimulated cell motility. This study indicated that the expression of both PAI-1 and C-FOS genes is necessary for wound-initiated normal rat kidney (NRK) epithelial cell migration [116].

While primary OPCs migrated toward the anode in the electrical field, neural stem cells and neural stem cell-derived OPCs (NSC-OPCs) migrated to the cathode in EFs [117, 118]. The activity of actin microfilaments at the leading edge of the cell determines the direction of cell migration. Polymerization of actin filaments underneath the plasma membrane is the main driving force for protrusions on the leading edge. The actin-related proteins 2 and 3 (ARP2/3) complex regulates the actin nucleation and is responsible for the formation of actin filament branches that can drive the lamellipodia protrusion. In contrast to the wild type of NSC-OPCs, those that were isolated from ARPC2$^{-/-}$ mouse embryos did not show cathodal migration in an EF. This study revealed that the ARP2/3 complex is required for the regulation of cathodal migration of NSC-OPCs in an EF [118].

References

1. Yang Y, Lewis R, Miller RH. Interactions between oligodendrocyte precursors control the onset of CNS myelination. Dev Biol. 2011;350(1):127–38.
2. Hughes EG, Appel B. The cell biology of CNS myelination. Curr Opin Neurobiol. 2016;39:93–100.
3. Baumann N, Pham-Dinh D. Biology of oligodendrocyte and myelin in the mammalian central nervous system. Physiol Rev. 2001;81(2):871–927.
4. Garcia-Verdugo JM, Doetsch F, Wichterle H, Lim DA, Alvarez-Buylla A. Architecture and cell types of the adult subventricular zone: in search of the stem cells. J Neurobiol. 1998;36(2):234–48.
5. Bunge MB, Bunge RP, Ris H. Ultrastructural study of remyelination in an experimental lesion in adult cat spinal cord. J Biophys Biochem Cytol. 1961;10:67–94.
6. Liu Y, Given KS, Harlow DE, Matschulat AM, Macklin WB, Bennett JL, et al. Myelin-specific multiple sclerosis antibodies cause complement-dependent oligodendrocyte loss and demyelination. Acta Neuropathol Commun. 2017;5(1):25.
7. Karttunen MJ, Czopka T, Goedhart M, Early JJ, Lyons DA. Regeneration of myelin sheaths of normal length and thickness in the zebrafish CNS correlates with growth of axons in caliber. PLoS One. 2017;12(5):e0178058.
8. Noble M, Murray K, Stroobant P, Waterfield MD, Riddle P. Platelet-derived growth factor promotes division and motility and inhibits premature differentiation of the oligodendrocyte/type-2 astrocyte progenitor cell. Nature. 1988;333(6173):560–2.
9. Richardson WD, Pringle N, Mosley MJ, Westermark B, Dubois-Dalcq M. A role for platelet-derived growth factor in normal gliogenesis in the central nervous system. Cell. 1988;53(2):309–19.
10. Horner PJ, Gage FH. Regenerating the damaged central nervous system. Nature. 2000;407(6807):963–70.
11. Yoon H, Radulovic M, Walters G, Paulsen AR, Drucker K, Starski P, et al. Protease activated receptor 2 controls myelin development, resiliency and repair. Glia. 2017;65(12):2070–86.
12. Privat A. Pathophysiology and treatment of spinal cord injury. Bulletin de l'Academie nationale de medecine. 2005;189(6):1109–17. discussion 17-8
13. Schwab JM, Brechtel K, Mueller CA, Failli V, Kaps HP, Tuli SK, et al. Experimental strategies to promote spinal cord regeneration—an integrative perspective. Prog Neurobiol. 2006;78(2):91–116.
14. Park E, Velumian AA, Fehlings MG. The role of excitotoxicity in secondary mechanisms of spinal cord injury: a review with an emphasis on the implications for white matter degeneration. J Neurotrauma. 2004;21(6):754–74.
15. Grossman SD, Rosenberg LJ, Wrathall JR. Temporal-spatial pattern of acute neuronal and glial loss after spinal cord contusion. Exp Neurol. 2001;168(2):273–82.
16. Crowe MJ, Bresnahan JC, Shuman SL, Masters JN, Beattie MS. Apoptosis and delayed degeneration after spinal cord injury in rats and monkeys. Nat Med. 1997;3(1):73–6.
17. Warden P, Bamber NI, Li H, Esposito A, Ahmad KA, Hsu CY, et al. Delayed glial cell death following wallerian degeneration in white matter tracts after spinal cord dorsal column cordotomy in adult rats. Exp Neurol. 2001;168(2):213–24.

18. Shuman SL, Bresnahan JC, Beattie MS. Apoptosis of microglia and oligodendrocytes after spinal cord contusion in rats. J Neurosci Res. 1997;50(5):798–808.
19. Barsoum MJ, Yuan H, Gerencser AA, Liot G, Kushnareva Y, Graber S, et al. Nitric oxide-induced mitochondrial fission is regulated by dynamin-related GTPases in neurons. EMBO J. 2006;25(16):3900–11.
20. Liot G, Bossy B, Lubitz S, Kushnareva Y, Sejbuk N, Bossy-Wetzel E. Complex II inhibition by 3-NP causes mitochondrial fragmentation and neuronal cell death via an NMDA- and ROS-dependent pathway. Cell Death Differ. 2009;16(6):899–909.
21. Ray SK, Matzelle DD, Wilford GG, Hogan EL, Banik NL. Inhibition of calpain-mediated apoptosis by E-64 d-reduced immediate early gene (IEG) expression and reactive astrogliosis in the lesion and penumbra following spinal cord injury in rats. Brain Res. 2001;916(1–2):115–26.
22. Ray SK, Hogan EL, Banik NL. Calpain in the pathophysiology of spinal cord injury: neuroprotection with calpain inhibitors. Brain Res Brain Res Rev. 2003;42(2):169–85.
23. Knoblach SM, Huang X, VanGelderen J, Calva-Cerqueira D, Faden AI. Selective caspase activation may contribute to neurological dysfunction after experimental spinal cord trauma. J Neurosci Res. 2005;80(3):369–80.
24. Brosius Lutz A, Barres BA. Contrasting the glial response to axon injury in the central and peripheral nervous systems. Dev Cell. 2014;28(1):7–17.
25. Kato K, Losada-Perez M, Hidalgo A. Gene network underlying the glial regenerative response to central nervous system injury. Dev Dynam. 2018;247(1):85–93.
26. Li Y, Zhang R, Hou X, Zhang Y, Ding F, Li F, et al. Microglia activation triggers oligodendrocyte precursor cells apoptosis via HSP60. Mol Med Rep. 2017;16(1):603–8.
27. Sellers DL, Maris DO, Horner PJ. Postinjury niches induce temporal shifts in progenitor fates to direct lesion repair after spinal cord injury. J Neurosci. 2009;29(20):6722–33.
28. Franklin RJ, Blakemore WF. Glial-cell transplantation and plasticity in the O-2A lineage—implications for CNS repair. Trends Neurosci. 1995;18(3):151–6.
29. Franklin RJ, Blakemore WF. To what extent is oligodendrocyte progenitor migration a limiting factor in the remyelination of multiple sclerosis lesions? Multi Scler (Houndmills, Basingstoke, England). 1997;3(2):84–7.
30. Blakemore WF, Franklin RJ. Transplantation options for therapeutic central nervous system remyelination. Cell Transplant. 2000;9(2):289–94.
31. Almad A, Sahinkaya FR, McTigue DM. Oligodendrocyte fate after spinal cord injury. Neurotherapeutics. 2011;8(2):262–73.
32. Shibuya S, Miyamoto O, Itano T, Mori S, Norimatsu H. Temporal progressive antigen expression in radial glia after contusive spinal cord injury in adult rats. Glia. 2003;42(2):172–83.
33. Zai LJ, Wrathall JR. Cell proliferation and replacement following contusive spinal cord injury. Glia. 2005;50(3):247–57.
34. Gilson J, Blakemore WF. Failure of remyelination in areas of demyelination produced in the spinal cord of old rats. Neuropathol Appl Neurobiol. 1993;19(2):173–81.
35. Guest JD, Hiester ED, Bunge RP. Demyelination and Schwann cell responses adjacent to injury epicenter cavities following chronic human spinal cord injury. Exp Neurol. 2005;192(2):384–93.
36. Akiyama Y, Honmou O, Kato T, Uede T, Hashi K, Kocsis JD. Transplantation of clonal neural precursor cells derived from adult human brain establishes functional peripheral myelin in the rat spinal cord. Exp Neurol. 2001;167(1):27–39.
37. Keirstead HS, Ben-Hur T, Rogister B, O'Leary MT, Dubois-Dalcq M, Blakemore WF. Polysialylated neural cell adhesion molecule-positive CNS precursors generate both oligodendrocytes and Schwann cells to remyelinate the CNS after transplantation. J Neurosci. 1999;19(17):7529–36.
38. Mujtaba T, Mayer-Proschel M, Rao MS. A common neural progenitor for the CNS and PNS. Dev Biol. 1998;200(1):1–15.
39. Tsai HH, Niu J, Munji R, Davalos D, Chang J, Zhang H, et al. Oligodendrocyte precursors migrate along vasculature in the developing nervous system. Science. 2016;351(6271):379–84.

References

40. Tripathi RB, Rivers LE, Young KM, Jamen F, Richardson WD. NG2 glia generate new oligodendrocytes but few astrocytes in a murine experimental autoimmune encephalomyelitis model of demyelinating disease. J Neurosci. 2010;30(48):16383–90.
41. Tsai HH, Frost E, To V, Robinson S, Ffrench-Constant C, Geertman R, et al. The chemokine receptor CXCR2 controls positioning of oligodendrocyte precursors in developing spinal cord by arresting their migration. Cell. 2002;110(3):373–83.
42. David S, Lacroix S. Molecular approaches to spinal cord repair. Annu Rev Neurosci. 2003;26:411–40.
43. Robinson S, Tani M, Strieter RM, Ransohoff RM, Miller RH. The chemokine growth-regulated oncogene-alpha promotes spinal cord oligodendrocyte precursor proliferation. J Neurosci. 1998;18(24):10457–63.
44. Wu Q, Miller RH, Ransohoff RM, Robinson S, Bu J, Nishiyama A. Elevated levels of the chemokine GRO-1 correlate with elevated oligodendrocyte progenitor proliferation in the jimpy mutant. J Neurosci. 2000;20(7):2609–17.
45. Popko B. Notch signaling: a rheostat regulating oligodendrocyte differentiation? Dev Cell. 2003;5(5):668–9.
46. Wang S, Sdrulla AD, diSibio G, Bush G, Nofziger D, Hicks C, et al. Notch receptor activation inhibits oligodendrocyte differentiation. Neuron. 1998;21(1):63–75.
47. Mogha A, D'Rozario M, Monk KR. G protein-coupled receptors in Myelinating glia. Trends Pharmacol Sci. 2016;37(11):977–87.
48. Doi T, Ogata T, Yamauchi J, Sawada Y, Tanaka S, Nagao M. Chd7 collaborates with Sox2 to regulate activation of Oligodendrocyte precursor cells after spinal cord injury. J Neurosci. 2017;37(43):10290–309.
49. Zhao Q, Han B, Wang Z, Gao C, Peng C, Shen J. Hollow chitosan-alginate multilayer microcapsules as drug delivery vehicle: doxorubicin loading and in vitro and in vivo studies. Nanomedicine. 2007;3(1):63–74.
50. He D, Marie C, Zhao C, Kim B, Wang J, Deng Y, et al. Chd7 cooperates with Sox10 and regulates the onset of CNS myelination and remyelination. Nat Neurosci. 2016;19(5):678–89.
51. Levine JM, Reynolds R. Activation and proliferation of endogenous oligodendrocyte precursor cells during ethidium bromide-induced demyelination. Exp Neurol. 1999;160(2):333–47.
52. Bambakidis NC, Miller RH. Transplantation of oligodendrocyte precursors and sonic hedgehog results in improved function and white matter sparing in the spinal cords of adult rats after contusion. The spine journal : official journal of the. N Am Spine Soc. 2004;4(1):16–26.
53. Wang L, Wei FX, Cen JS, Ping SN, Li ZQ, Chen NN, et al. Early administration of tumor necrosis factor-alpha antagonist promotes survival of transplanted neural stem cells and axon myelination after spinal cord injury in rats. Brain Res. 2014;1575:87–100.
54. Windrem MS, Nunes MC, Rashbaum WK, Schwartz TH, Goodman RA, McKhann G 2nd, et al. Fetal and adult human oligodendrocyte progenitor cell isolates myelinate the congenitally dysmyelinated brain. Nat Med. 2004;10(1):93–7.
55. Seilhean D, Gansmuller A, Baron-Van Evercooren A, Gumpel M, Lachapelle F. Myelination by transplanted human and mouse central nervous system tissue after long-term cryopreservation. Acta Neuropathol. 1996;91(1):82–8.
56. Cao Q, Xu XM, Devries WH, Enzmann GU, Ping P, Tsoulfas P, et al. Functional recovery in traumatic spinal cord injury after transplantation of multineurotrophin-expressing glial-restricted precursor cells. J Neurosci. 2005;25(30):6947–57.
57. Yang J, Xiong LL, Wang YC, He X, Jiang L, Fu SJ, et al. Oligodendrocyte precursor cell transplantation promotes functional recovery following contusive spinal cord injury in rats and is associated with altered microRNA expression. Mol Med Rep. 2018;17(1):771–82.
58. Garcia-Alias G, Lopez-Vales R, Fores J, Navarro X, Verdu E. Acute transplantation of olfactory ensheathing cells or Schwann cells promotes recovery after spinal cord injury in the rat. J Neurosci Res. 2004;75(5):632–41.
59. Munoz-Quiles C, Santos-Benito FF, Llamusi MB, Ramon-Cueto A. Chronic spinal injury repair by olfactory bulb ensheathing glia and feasibility for autologous therapy. J Neuropathol Exp Neurol. 2009;68(12):1294–308.

60. Erceg S, Ronaghi M, Oria M, Rosello MG, Arago MA, Lopez MG, et al. Transplanted oligodendrocytes and motoneuron progenitors generated from human embryonic stem cells promote locomotor recovery after spinal cord transection. Stem Cells (Dayton, Ohio). 2010;28(9):1541–9.
61. Nistor GI, Totoiu MO, Haque N, Carpenter MK, Keirstead HS. Human embryonic stem cells differentiate into oligodendrocytes in high purity and myelinate after spinal cord transplantation. Glia. 2005;49(3):385–96.
62. Keirstead HS, Nistor G, Bernal G, Totoiu M, Cloutier F, Sharp K, et al. Human embryonic stem cell-derived oligodendrocyte progenitor cell transplants remyelinate and restore locomotion after spinal cord injury. J Neurosci. 2005;25(19):4694–705.
63. Mackay-Sim A, Feron F, Cochrane J, Bassingthwaighte L, Bayliss C, Davies W, et al. Autologous olfactory ensheathing cell transplantation in human paraplegia: a 3-year clinical trial. Brain. 2008;131.(Pt 9:2376–86.
64. Lima C, Pratas-Vital J, Escada P, Hasse-Ferreira A, Capucho C, Peduzzi JD. Olfactory mucosa autografts in human spinal cord injury: a pilot clinical study. J Spinal Cord Med. 2006;29(3):191–203. discussion 4-6
65. Chhabra HS, Lima C, Sachdeva S, Mittal A, Nigam V, Chaturvedi D, et al. Autologous olfactory [corrected] mucosal transplant in chronic spinal cord injury: an Indian pilot study. Spinal Cord. 2009;47(12):887–95.
66. Huang H, Wang H, Chen L, Gu Z, Zhang J, Zhang F, et al. Influence factors for functional improvement after olfactory ensheathing cell transplantation for chronic spinal cord injury. Zhongguo xiu fu chong jian wai ke za zhi = Zhongguo xiufu chongjian waike zazhi = Chin J Reparat Reconst Surg. 2006;20(4):434–8.
67. Spilker MH, Yannas IV, Kostyk SK, Norregaard TV, Hsu HP, Spector M. The effects of tubulation on healing and scar formation after transection of the adult rat spinal cord. Restor Neurol Neurosci. 2001;18(1):23–38.
68. King VR, Alovskaya A, Wei DY, Brown RA, Priestley JV. The use of injectable forms of fibrin and fibronectin to support axonal ingrowth after spinal cord injury. Biomaterials. 2010;31(15):4447–56.
69. Li X, Yang Z, Zhang A, Wang T, Chen W. Repair of thoracic spinal cord injury by chitosan tube implantation in adult rats. Biomaterials. 2009;30(6):1121–32.
70. Hassan W, Dong Y, Wang W. Encapsulation and 3D culture of human adipose-derived stem cells in an in-situ crosslinked hybrid hydrogel composed of PEG-based hyperbranched copolymer and hyaluronic acid. Stem Cell Res Ther. 2013;4(2):32.
71. Mekhail M, Almazan G, Tabrizian M. Purine-crosslinked injectable chitosan sponges promote oligodendrocyte progenitor cells' attachment and differentiation. Biomater Sci. 2015;3(2):279–87.
72. Kim YT, Caldwell JM, Bellamkonda RV. Nanoparticle-mediated local delivery of methylprednisolone after spinal cord injury. Biomaterials. 2009;30(13):2582–90.
73. Ranjan OP, Shavi GV, Nayak UY, Arumugam K, Averineni RK, Meka SR, et al. Controlled release chitosan microspheres of mirtazapine: in vitro and in vivo evaluation. Arch Pharm Res. 2011;34(11):1919–29.
74. Mathew ST, Devi SG, Prasanth VV, Vinod B. Formulation and in vitro-in vivo evaluation of ketoprofen-loaded albumin microspheres for intramuscular administration. J Microencapsul. 2009;26(5):456–69.
75. Nagai N, Kumasaka N, Kawashima T, Kaji H, Nishizawa M, Abe T. Preparation and characterization of collagen microspheres for sustained release of VEGF. J Mater Sci Mater Med. 2010;21(6):1891–8.
76. Wood MD, Kim H, Bilbily A, Kemp SW, Lafontaine C, Gordon T, et al. GDNF released from microspheres enhances nerve regeneration after delayed repair. Muscle Nerve. 2012;46(1):122–4.
77. Yao R, Zhang R, Luan J, Lin F. Alginate and alginate/gelatin microspheres for human adipose-derived stem cell encapsulation and differentiation. Biofabrication. 2012;4(2):025007.

References

78. Natesan S, Baer DG, Walters TJ, Babu M, Christy RJ. Adipose-derived stem cell delivery into collagen gels using chitosan microspheres. Tissue Eng Part A. 2010;16(4):1369–84.
79. Chan BP, Hui TY, Wong MY, Yip KH, Chan GC. Mesenchymal stem cell-encapsulated collagen microspheres for bone tissue engineering. Tissue Eng Part C Methods. 2010;16(2):225–35.
80. Lim F, Sun AM. Microencapsulated islets as bioartificial endocrine pancreas. Science. 1980;210(4472):908–10.
81. Ricci M, Blasi P, Giovagnoli S, Rossi C, Macchiarulo G, Luca G, et al. Ketoprofen controlled release from composite microcapsules for cell encapsulation: effect on post-transplant acute inflammation. J Controlled Release. 2005;107(3):395–407.
82. Yao L, Phan F, Li Y. Collagen microsphere serving as a cell carrier supports oligodendrocyte progenitor cell growth and differentiation for neurite myelination in vitro. Stem Cell Res Ther. 2013;4(5):109.
83. Berndt M, Li Y, Seyedhassantehrani N, Yao L. Fabrication and characterization of microspheres encapsulating astrocytes for neural regeneration. ACS Biomater Sci Eng. 2017;3(7):1313–21.
84. Yao L, O'Brien N, Windebank A, Pandit A. Orienting neurite growth in electrospun fibrous neural conduits. J Biomed Mater Res B Appl Biomater. 2009;90((2):483–91.
85. Zander NE, Orlicki JA, Rawlett AM, Beebe TP Jr. Surface-modified nanofibrous biomaterial bridge for the enhancement and control of neurite outgrowth. Biointerphases. 2010;5(4):149–58.
86. Prabhakaran MP, Venugopal JR, Chyan TT, Hai LB, Chan CK, Lim AY, et al. Electrospun biocomposite nanofibrous scaffolds for neural tissue engineering. Tissue Eng Part A. 2008;14(11):1787–97.
87. Zamani F, Amani-Tehran M, Latifi M, Shokrgozar MA, Zaminy A. Promotion of spinal cord axon regeneration by 3D nanofibrous core-sheath scaffolds. J Biomed Mater Res A. 2014;102(2):506–13.
88. Wang W, Itoh S, Konno K, Kikkawa T, Ichinose S, Sakai K, et al. Effects of Schwann cell alignment along the oriented electrospun chitosan nanofibers on nerve regeneration. J Biomed Mater Res A. 2009;91((4):994–1005.
89. Neal RA, Tholpady SS, Foley PL, Swami N, Ogle RC, Botchwey EA. Alignment and composition of laminin-polycaprolactone nanofiber blends enhance peripheral nerve regeneration. J Biomed Mater Res A. 2012;100((2):406–23.
90. Yu W, Zhao W, Zhu C, Zhang X, Ye D, Zhang W, et al. Sciatic nerve regeneration in rats by a promising electrospun collagen/poly(epsilon-caprolactone) nerve conduit with tailored degradation rate. BMC Neurosci. 2011;12:68.
91. Chan JR, Watkins TA, Cosgaya JM, Zhang C, Chen L, Reichardt LF, et al. NGF controls axonal receptivity to myelination by Schwann cells or oligodendrocytes. Neuron. 2004;43(2):183–91.
92. Wang H, Tewari A, Einheber S, Salzer JL, Melendez-Vasquez CV. Myosin II has distinct functions in PNS and CNS myelin sheath formation. J Cell Biol. 2008;182(6):1171–84.
93. O'Meara RW, Ryan SD, Colognato H, Kothary R. Derivation of enriched oligodendrocyte cultures and oligodendrocyte/neuron myelinating co-cultures from post-natal murine tissues. J Visual Exp. 2011;(54)
94. Lee S, Chong SY, Tuck SJ, Corey JM, Chan JR. A rapid and reproducible assay for modeling myelination by oligodendrocytes using engineered nanofibers. Nat Protoc. 2013;8(4):771–82.
95. Lee S, Leach MK, Redmond SA, Chong SY, Mellon SH, Tuck SJ, et al. A culture system to study oligodendrocyte myelination processes using engineered nanofibers. Nat Methods. 2012;9(9):917–22.
96. Li Y, Ceylan M, Shrestha B, Wang H, Lu QR, Asmatulu R, et al. Nanofibers support oligodendrocyte precursor cell growth and function as a neuron-free model for myelination study. Biomacromolecules. 2014;15(1):319–26.
97. Diao HJ, Low WC, Milbreta U, Lu QR, Chew SY. Nanofiber-mediated microRNA delivery to enhance differentiation and maturation of oligodendroglial precursor cells. J Controlled Release. 2015;208:85–92.

98. Lee HU, Blasiak A, Agrawal DR, Loong DTB, Thakor NV, All AH, et al. Subcellular electrical stimulation of neurons enhances the myelination of axons by oligodendrocytes. PLoS One. 2017;12(7):e0179642.
99. Makinodan M, Rosen KM, Ito S, Corfas G. A critical period for social experience-dependent oligodendrocyte maturation and myelination. Science. 2012;337(6100):1357–60.
100. Liu J, Dietz K, DeLoyht JM, Pedre X, Kelkar D, Kaur J, et al. Impaired adult myelination in the prefrontal cortex of socially isolated mice. Nat Neurosci. 2012;15(12):1621–3.
101. Li Q, Houdayer T, Liu S, Belegu V. Induced neural activity promotes an oligodendroglia regenerative response in the injured spinal cord and improves motor function after spinal cord injury. J Neurotrauma. 2017;34(24):3351–61.
102. McKasson MJ, Huang L, Robinson KR. Chick embryonic Schwann cells migrate anodally in small electrical fields. Exp Neurol. 2008;211(2):585–7.
103. Yao L, Li Y, Knapp J, Smith P. Exploration of molecular pathways mediating electric field-directed Schwann cell migration by RNA-seq. J Cell Physiol. 2015;230(7):1515–24.
104. Li Y, Weiss M, Yao L. Directed migration of embryonic stem cell-derived neural cells in an applied electric field. Stem Cell Rev. 2014;10(5):653–62.
105. Zhang J, Calafiore M, Zeng Q, Zhang X, Huang Y, Li RA, et al. Electrically guiding migration of human induced pluripotent stem cells. Stem Cell Rev. 2011;7(4):987–96.
106. Hirotsu T, Saeki S, Yamamoto M, Iino Y. The Ras-MAPK pathway is important for olfaction in Caenorhabditis elegans. Nature. 2000;404(6775):289–93.
107. Nuccitelli R, Smart T. Extracellular calcium levels strongly influence neural crest cell Galvanotaxis. Biol Bull. 1989;176(2s):130–5.
108. Zhao M, Pu J, Forrester JV, McCaig CD. Membrane lipids, EGF receptors, and intracellular signals colocalize and are polarized in epithelial cells moving directionally in a physiological electric field. FASEB J. 2002;16(8):857–9.
109. Ji J, Liu R, Tong T, Song Y, Jin S, Wu M, et al. Gadd45a regulates beta-catenin distribution and maintains cell-cell adhesion/contact. Oncogene. 2007;26(44):6396–405.
110. Shan Z, Li G, Zhan Q, Li D. Gadd45a inhibits cell migration and invasion by altering the global RNA expression. Cancer Biol Ther. 2012;13(11):1112–22.
111. Alexopoulou AN, Leao M, Caballero OL, Da Silva L, Reid L, Lakhani SR, et al. Dissecting the transcriptional networks underlying breast cancer: NR4A1 reduces the migration of normal and breast cancer cell lines. Breast Cancer Res. 2010;12(4):R51.
112. Kanse SM, Kost C, Wilhelm OG, Andreasen PA, Preissner KT. The urokinase receptor is a major vitronectin-binding protein on endothelial cells. Exp Cell Res. 1996;224(2):344–53.
113. Kjoller L, Kanse SM, Kirkegaard T, Rodenburg KW, Ronne E, Goodman SL, et al. Plasminogen activator inhibitor-1 represses integrin- and vitronectin-mediated cell migration independently of its function as an inhibitor of plasminogen activation. Exp Cell Res. 1997;232(2):420–9.
114. Loskutoff DJ, Curriden SA, Hu G, Deng G. Regulation of cell adhesion by PAI-1. APMIS. 1999;107(1):54–61.
115. Stefansson S, Lawrence DA. The serpin PAI-1 inhibits cell migration by blocking integrin alpha V beta 3 binding to vitronectin. Nature. 1996;383(6599):441–3.
116. Kutz SM, Providence KM, Higgins PJ. Antisense targeting of c-fos transcripts inhibits serum- and TGF-beta 1-stimulated PAI-1 gene expression and directed motility in renal epithelial cells. Cell Motil Cytoskeleton. 2001;48(3):163–74.
117. Li L, El-Hayek YH, Liu B, Chen Y, Gomez E, Wu X, et al. Direct-current electrical field guides neuronal stem/progenitor cell migration. Stem cells (Dayton, Ohio). 2008;26(8):2193–200.
118. Li Y, Wang PS, Lucas G, Li R, Yao L. ARP2/3 complex is required for directional migration of neural stem cell-derived oligodendrocyte precursors in electric fields. Stem Cell Res Ther. 2015;6:41.

Chapter 3
Application of Schwann Cells in Neural Tissue Engineering

Li Yao and Priyanka Priyadarshani

Abstract In the peripheral nervous system (PNS), Schwann cells (SCs) are the principal glial cells that myelinate axons. Unlike the central nervous system (CNS), the peripheral nervous system has the potential to regenerate after injury. SCs effectively respond to injury and help in axon regeneration during the early stage of peripheral nerve injury. Schwann cells also participate in the remyelination of axons in spinal cord injury (SCI). Following SCI, endogenous SCs migrate and invade the injury site where they associate with regenerating axons. Recent studies have demonstrated that Schwann cell transplantation can significantly enhance regeneration post-neural tissue injury. In this chapter, we review the critical role of Schwann cell in peripheral and spinal cord injuries.

Keywords Spinal cord · Vascularization · Extracorporeal shock wave therapy · Hypothermia · Biomaterial scaffolds · Reconstruction of vascular structure · Molecular therapy · Ischemia · Growth factors

3.1 Origin of Schwann Cells

Schwann cells (SCs) are the principal glial cells in the peripheral nervous system (PNS) [1] that are tightly associated with neuronal axons during early stages of embryonic development and are capable of migrating long distances along the nerves. They are descendants of neural crest cells (NCCs), which during early divisions, migrate toward the ventral pathway and differentiate into neurons and Schwann cell precursors (SCPs). During early embryonic phases, NCCs break away from the dorsal aspect of the neural tube and differentiate to form a variety of cell types, such as neurons, glial cells, pigment cells, cartilage, and smooth muscle [2]. SCPs emerge from specified NCCs that represent the first transitional stage in the Schwann cell lineage [2]. SCPs precede and develop into immature SCs that are randomly associated with mixed bundles of large and small caliber axons [3]. These random associations of immature SCs with axons determine the further specification of immature SCs [4]. In the PNS, SCs and axons are enveloped in a specialized form of extracellular matrices (ECMs) called basal lamina [1]. SCs are tightly associated with axons and thus distinguish the axons into myelinating nerve fiber and nonmyelinating nerve fiber. Both types of nerve fiber play a significant role in maintaining the PNS function [1]. Myelinating nerves consist of a single axon and SCs ensheathing axons larger than 1 μm, while nonmyelinating fibers are small caliber axons smaller than 1 μm and form Remak bundles [1, 5–7]. Thus, SC progression includes three main transition steps: transformation of NCCs to SCPs, then SCPs to immature SCs, and then immature SCs to myelin and non-myelin forming mature SCs. Each stage of SC development is tightly regulated by transcription factors [8]. One of the central molecules for SC development and regulation is transcription factor SOX-10. It has been revealed that SOX-10 plays an integral role in SC specification and development through collaboration with other transcription factors [9]. A number of signals, including the neuregulin 1 (NRG1) gene, endothelins, and the Notch signaling pathway, regulate the differentiation of SCPs to SCs [3]. Other promising regulatory factors, POU-domain factors Oct-6 (also known as POU3F1, SCIP, and Tst-1) and the zinc-finger protein, early growth response 2 (EGR2, also known as Krox-20), are required in SC development [10]. SOX-10, Oct-6, and Krox-20 interact with each other in association with other transcription factors and cofactors such as NFAT, SREBP, SOX-2, c-Jun, Nab, HDAC, and others to activate gene expression and to control terminal differentiation [8]. In addition to transcriptional activators factors, a transcriptional repression factor, zinc-finger protein ZEB2 (also known as SIP1), is also important for differentiation and myelination during SC development. The absence of ZEB2 in mouse leads to the failure of axonal sorting, the virtual absence of myelin membranes, and the continuous repression of lineage progression in SCs [11, 12]. Overall, each developmental stage in the Schwann cell lineage is regulated by different sets of molecules, signaling responses, and tissue association [3].

3.2 Peripheral Nerve Injury and Role of Schwann Cells in Nerve Regeneration

Nerve injury causes long-term disability, resulting in a lower quality of life and morbidity. Nerve injury occurs whenever there is damage in the brain, spinal cord, and peripheral nerves located throughout the body. Injuries caused by collisions, fractures, vehicle accidents, and gunshot wounds are the primary causes of nerve injury. Moreover, diabetes, cancer, and surgery can also cause injury [13]. The neuropathic pain and symptoms depend upon the location and type of nerves that are affected.

The peripheral nervous system has the potential to regenerate after injury, unlike the central nervous system (CNS) [14]. This is because, following injury, while myelin debris is readily phagocytosed and cleared in the PNS, glia scarring blocks axonal growth, and debris remains to produce inhibitory molecules that restrict regeneration in the CNS [15]. In the process of regeneration, the proximal nerve stump of the damaged nerve is pulled back to the node of Ranvier, and the distal nerve stump undergoes Wallerian degeneration. This axonal degeneration is followed by degradation of the myelin sheath. The macrophages and Schwann cells serve to clear the debris from the degeneration [16, 17]. During the regenerative state following peripheral nerve injury, signals from the injured site are delivered to the soma, which activates the cell survival pathways and up-regulates numerous regeneration-associated genes. These sequential events following injury include neurochemical changes, functional alterations in excitability, growth cone reconstruction, local protein synthesis, and activation of multiple signaling pathways in axonal regeneration [18]. SCs are the crucial component in the regeneration because they provide mechanical strength that is essential for PNS mechanoprotection as well as mechanosensitivity, which are critical for PNS development and myelination [1]. SCs envelop peripheral axons with a spirally wrapped myelin sheath [19]. During nerve injury, SCs down-regulate myelin protein production and up-regulate the expression of growth-promoting genes, including cell adhesion molecules, growth factors, laminin, fibronectin, heparin sulfate, and collagen [20–22]. SCs secrete neurotrophins, which play a vital role in peripheral regeneration [18]. They also form Bungner bands, which aid and serve as a scaffold for axon regeneration [23, 24]. Various mRNA transcription factors regulate axon regeneration and neuronal survival after injury [25]. Eighteen to 28 h post-peripheral injury, 26 transcription factor families are associated with changes in gene expression. These families include the signal transducer and activator of transcription (STAT), hepatocyte nuclear factor (HNF), upstream stimulatory factor (USF), Jun, Smad, SRF, and ERα [25]. Six hours post-peripheral nerve injury, proteins such as ATF3, neural peptide Y, Arginase I, and ankyrin repeat domain 1 (ANKRD1) coordinate to regenerate the injured nerve [26]. Various other proteins like axonal regeneration neurotrophic factors, cytokines, GP-43, tubulin, actin, neuropeptides, and multiple adhesion molecules assist in axonal regeneration. In addition to these factors, specified targeted factors guide sensory and motor axons to their specific end-organ target [27, 28]. It takes about 3–4 weeks to restore the structure and functions of injured neurons. The disruption in any of these processes could impair regeneration [29]. Although SCs effectively remyelinate axons and help in axon regeneration from the injured proximal nerve end into the distal nerve end during the early stage, cellular and molecular changes that initially assist axonal regeneration fail to promote regeneration a few weeks after injury. Furthermore, inhibitory molecules from debris gradually fill the nerve pathway, thereby causing restriction in the regeneration [30, 31]. Therefore, therapeutic methods that can support active nerve regeneration for the long-term repair process in an injury having a long nerve gap should be developed. Recent study revealed that histone deacetylase 3 (HDAC3; a histone-modifying enzyme) is a potent inhibitor of peripheral myelinogenesis. Inhibition of HDAC3 enhanced myelin growth and regeneration and improved functional recovery after peripheral nerve injury in mice. Myelin thickness in sciatic nerves increased after Schwann cell-specific deletion of Hdac3 in mice [32].

3.3 Autologous Nerve Grafting and Schwann Cell Transplantation for Nerve Repair

Autologous nerve grafting for peripheral nerve repair is a widely practiced technique. It is used to mechanically guide the axonal outgrowth toward the distal nerve stump. A sural nerve is commonly used for grafting purposes because of its superficial location and suitable length for nerve repair [33]. In the recent treatment of a human patient, autologous SCs in combination with a sural nerve graft were used to repair a 7.5 cm sciatic nerve defect (Table 3.1). A sural nerve biopsy and peripheral nerve tissue were obtained from the traumatized sciatic nerve stumps of the human patient. In this study, human SCs were extracted by enzymatic digestion of the peripheral nerve segment and cultured in the presence of heregulin-beta 1 and forskolin. After culturing for two passages, the SCs were transplanted back into the long-segment (7.5 cm) sciatic nerve defect with the sural nerve autograft. Continuous nerve repair and recovery of the sensory and motor neurons were confirmed. Moreover, no tumor formation was seen after this treatment [34]. SC support in axon regeneration can last for a short period of time, which is not sufficient for the recovery of nerve injury with a long nerve gap. In order to overcome this problem, a new technique has been developed, whereby short nerve grafts were inserted side by side between the intact donor nerve and the denervated recipient nerve, which assisted as a bridge for the donor axons to regenerate into the recipient nerve. This study showed that denervated SCs in the bridge and the recipient denervated nerve stump dedifferentiated into a proliferative, nonmyelinating phenotype. As the donor axons grew across the nerve bridges and into the denervated CP nerve, the SCs redifferentiated into the myelinating phenotype [35]. This case shows the clinical feasibility of a nerve graft. In a similar study of a delayed rat nerve-repair model, side-by-side cross bridging was performed by placing nerve autografts at right angles between a donor tibial (TIB) nerve and a recipient denervated common peroneal (CP) nerve. These side-by-side cross bridges ameliorated poor regeneration after delayed nerve repair by sustaining the growth-permissive state of the denervated nerve stumps [36]. Recently, an optimized allograft was fabricated by decellularizing rat nerves using chemicals, enzymes, and irradiation. The processed nerve segments were then stored at either 4 °C or –80 °C. Results showed that the ultrastructure of the nerve segments can be preserved with all decellularization protocols when the nerve segments are stored at 4 °C. Also, reduced immunogenicity, diminished cellular debris, and SC elimination was observed when elastase was used in the decellularization process [37]. This optimized technique can be further utilized for nerve-regeneration purposes.

3.3 Autologous Nerve Grafting and Schwann Cell Transplantation for Nerve Repair   41

Table 3.1 Grafting of Schwann cell for neural regeneration

Cell type	Treatment	Animal model	Tissue	Procedure	Results	Refs.
SCs	N/A	Human	Nerve	Cultured SCs were combined with a sural nerve graft to bridge a 7.5-cm-long sciatic nerve defect.	Sensory and motor function recovery, and absence of tumor formation were observed.	[34]
SCs	N/A	Rat	Nerve	Three nerve bridges were placed side by side between an intact donor tibial (TIB) nerve and a recipient denervated common peroneal (CP) distal nerve stump.	Regeneration and myelination patterns of donor axon in the denervated recipient were observed.	[35]
SCs precursors derived from neural crest cells resident in bone marrow (BM-NCCs)	N/A	Rat	Nerve	Rat sciatic nerve was crushed, followed by injection with BM-NCCs or BM-NCCs-derived SCs.	Group treated with BM-NCCs yielded a better outcome of nerve regeneration and function restoration than BM-NCCs-derived SCs.	[104]
SCs	Frankincense extract	Rat	Nerve	In vivo: Rat sciatic nerve was crushed and injected with frankincense extract. In vitro: SCs were cultured with medium containing frankincense extract.	Frankincense extract improved sciatic nerve regeneration and promoted the function of a crushed sciatic nerve in vivo and enhance the proliferation of SCs in vitro.	[38]
Human SCs and rat SCs	BDNF	In vitro	N/A	Human and rat SCs were cultured and treated with BDNF.	BDNF activated the JAK/STAT pathway in SCs that resulted in secretion of cytokines such as IL-6 and OSM-M.	[49]

(continued)

Table 3.1 (continued)

Cell type	Treatment	Animal model	Tissue	Procedure	Results	Refs.
SCs	Retinoic acid (RA)	In vitro	N/A	SCs were seeded on an Ibidi culture insert and placed onto PLL-treated well plates. After 24 hours of cell culture, the cells were transfected with NEDD9-specific siRNA. RA was added next to the transfected cells.	RA increased SC migration and could be a major regulator of SC migration after nerve injury.	[105]
SCs	Metformin	In vitro	N/A	SCs were introduced to a hypoxia condition prior to being supplemented with metformin.	Metformin helps SCs recover from hypoxia injury and maintain biological activities, such as increasing the expression and secretion of BDNF, NGF, GDNF, and NCAM.	[39]
SCs	N/A	Rat	N/A	The expression of SAM68 and its biological significance was studied in a sciatic nerve crush.	SAM68 may participate in SC proliferation partially via the PI3K/Akt pathway and also may regulate regeneration after a sciatic nerve crush.	[106]
SCs	miR-sc3	In vitro	N/A	Cultured SCs were transfected with synthetic miR-sc3 and miR-sc3 inhibitor.	Overexpression and silencing of miR-sc3 promoted and inhibited SC proliferation and migration, respectively.	[107]
SCs and OEG	N/A	Rat	Spinal cord	The injured spinal cord was transplanted with cultured adult rat SCs or olfactory ensheathing glia cells (OEGCs).	An SC graft is more efficient in promoting axonal regeneration than OEG.	[98]

(continued)

3.3 Autologous Nerve Grafting and Schwann Cell Transplantation for Nerve Repair

Table 3.1 (continued)

Cell type	Treatment	Animal model	Tissue	Procedure	Results	Refs.
SCs	Edaravone	Rat	Spinal cord	SCs extracted from a rat sciatic nerve were transplanted into the trisected rat spinal cord. Simultaneously after transplantation, edaravone was injected through the caudal vein.	SCs survived and migrated to the center of the spinal cord injury region in rats after combined treatment with edaravone and SCs. Improved motor function and neurophysiological function were observed.	[95]
SCs	Polyamine putrescine	Rat	Spinal cord	A rat moderate spinal cord injury received SC transplantation, SC transplantation with acute putrescine administration (30 minutes after injury), or SC transplantation with putrescine administration (after 1 week of injury).	Administration of putrescine and SC implantation 1 week after injury promoted growth of the grafted SCs, and enhanced axonal (sensory and serotonergic) sparing and/or growth and functional recovery.	[96]
SCs	Human telomerase reverse transcriptase (hTERT)	Rat	Spinal cord	SCs without hTERT transfection or SCs with hTERT transfection were grafted into a rat with SCI.	Transplantation of SCs with hTERT transfection in an injured spinal cord showed fewer apoptotic cells, improved tissue repair, increased number of PKH-26-positive cells at the injury site, and improvement in the recovery of lower limb motor function, compared to the group treated with non-transfected SCs.	[99]

3.4 Synthetic and Biological Molecules for Promoting Nerve Repair

A number of synthetic and biological molecules have been used to improve nerve regeneration and functional recovery. The frankincense extract, a fragrant gum from trees of the genus *Boswellia*, was commonly used for swelling, inflammatory diseases, tumor, and blood circulation stimulation. One study tested the effect of frankincense extract on peripheral nerve regeneration. In a rat nerve crush injury model, the wounded nerve was treated with a high dose of frankincense extract, and results showed promoted regeneration of the sciatic nerve, increased expression of GAP-43, and enhanced proliferation of SCs [38]. Metformin, which is usually used for the treatment of diabetes mellitus type-2, has been proven to generate beneficial effects on nerve regeneration and to maintain biological activity of SCs [39]. Herceptin, a monoclonal antibody, was administered to transected Sprague-Dawley rats immediately after injury or 4 months after chronic denervation. This study demonstrated that Herceptin administration accelerated the rate of motor and sensory neuron regeneration, increased the quantity of SCs in the distal stump after the first week, and increased the number of myelinated axons regenerated after 4 weeks of repair [40]. An in vitro study was done to investigate the molecular mechanism of protocatechuic acid (PCA), a dihydroxy benzoic acid. RSC96 SCs treated with PCA showed enhanced proliferation due to the activation of IGF-IR-Akt pathway. PCA regulates the cell cycle and promotes the anti-apoptotic proteins [41]. The combination of a neuroprotective reagent Salidroside (SDS) and poly (lactic-co-glycolic acid) (PLGA) was implanted into a 12-mm sciatic nerve gap of a rat. This study showed enhanced proliferation of SCs, and functional and morphological improvements of the regenerated nerve [42].

Growth factors can effectively ameliorate axonal growth in nerve regeneration [43]. Saceda et al. repaired a transected rat ulnar nerve by suturing a silastic tube to the proximal and distal nerve ends. This study showed that rats receiving systematic delivery of a growth hormone (GH) enhanced their recovery with increased nerve fiber density and myelination and reduced scarring [44]. Similarly, in another experiment, rats received GH treatment after sciatic nerve transection and epineurial suture repair. These GH-treated rats showed increased cellular density, myelination, and immense axons and SCs. Additionally, the GH-treated group showed a higher compound muscle action potential (CMAP) amplitude compared with the non-treated group [45]. In another study, the sciatic nerve was transected and repaired while the femoral nerve was transected without repair. The animals in this study were treated with GH for 5 weeks, and then the axonal regeneration was assessed. Animals receiving GH showed increased body mass, an increased number of regenerating myelinated axons, reduced muscle atrophy, and enhanced muscle reinnervation, compared with the non-treated control animals [46]. Brain-derived neurotrophic factor (BDNF) was also found to influence the growth and regenerative capacity of neurons [47]. BDNF is generated by SCs after peripheral nerve injury, which leads to its elevation in injured nerve and thus helps to prevent neuronal death, promotes neuronal activity, and enhances axonal growth [48]. Four different cell lines—human neuroblastoma BE(2)-C, SH-SY5Y, human SCs, and rat SCs RT4-D6P2T—were treated with BDNF to study the activation of the JAK/STAT pathway. This study showed that treatment with BDNF activated the JAK/STAT pathway in both human and rat SCs, resulting in the secretion of cytokines (such as IL-6 and OSM-M), which facilitate axonal regeneration, but the JAK/STAT pathway was not activated in the human neuroblastoma cells. This study indicated that BDNF-induced cytokine production in SCs can improve nerve regeneration [49].

Gene therapy is potential therapeutic strategy to promote nerve regeneration. SCs can be genetically modified to express proteins that can assist in nerve regeneration. Gene vectors can also be directly delivered to the injured nerves or cells to generate therapeutic molecules that promote nerve regeneration [50]. It has been demonstrated that Kruppel-like Factor 7 (KLF7) is a key transcription factor strongly activated during nerve injury [51, 52]. KLF7 can primarily enhance axonal regeneration of the peripheral motor and sensory nerve [53]. Because KLF7 is considered to be a stimulating factor for axonal regeneration, a study was carried out to overexpress KLF in an SC graft that could accelerate sciatic nerve regeneration. In this experiment, C57BL/6 mice were allografted by an acellular nerve (ANA) with either an injection of cell culture medium, SCs, or AAV2-KLF7-transfected SCs. Results showed that KLF7-overexpressing SCs stimulated axonal regeneration, myelination, and functional recovery of the injured peripheral nerve [54]. Another in vitro study showed that transfection of cultured SCs with SAM68 or miR-sc3 ameliorate cell proliferation, differentiation, and migration. Similarly, glial cell line-derived neurotrophic factor (GDNF) has proven to be useful in promoting axon survival and regeneration. It was demonstrated that transplantation of SCs transduced with a tetracycline-inducible (Tet-On) GDNF overexpressing lentivirus promoted nerve regeneration in a long nerve defect model. However, uncontrolled GDNF expression can actually impair regeneration. So careful spatial and temporal control of GDNF delivery is required in this type of treatment [55].

3.5 Stem Cell-Derived Schwann Cells for Neural Regeneration

SCs derived from human embryonic stem cells (hESCs) provide an alternative source of SCs for neural repair. Studies were performed to test the developmental process and biological functions of these cells in nerve regeneration. The differentiated SCs showed close association with axons, thereby indicating that these cells can be a possible source of SCs for nerve repair [56]. In one study, mouse ESCs were induced and differentiated into neural cells, which were then transplanted into the gap of transected rat sciatic nerve. Three months post-surgery, the neural cells derived from ESCs survived and differentiated into myelinating cells [57]. In one study, a three-dimensional sphere of cells was generated by human embryonic stem cell-derived mesenchymal stem cells (hESC-MSCs). The hESC-MSCs spheres were transplanted into the mice with a 2-mm gap of transected sciatic nerve. Results showed almost complete functional recovery of the injured peripheral nerve [58]. In another study, an acellular nerve allograft (ANA) injected with allogenic skin-derived precursor differentiated Schwann cells (SKP-SCs) and heregulin-beta 1 was grafted into rats having a 15-mm lesion in the sciatic nerve, which enhanced sciatic nerve function and promoted peripheral nerve repair [59].

The efficiency of different growth factors for SCs differentiation from stem cells was studied in one previous study. Three types of growth factors—basic fibroblast growth factor (bFGF), nerve growth factor (NGF), and neuregulin-1 beta—were investigated for their potential effect to induce differentiation of neural stem cells (NSCs) toward SCs. The NSCs isolated from rat dorsal root ganglion cells (DRGs) were separately treated with these growth factors. Results showed that bFGF has the highest potency to differentiate DRG-NSCs toward SC-like cells. Thus, this helps to establish an effective method to obtain an increase in yield of SCs from NSCs [60].

3.6 Schwann Cells and Biomaterial for Neural Regeneration

Due to the limitation in nerve tissue and available nerve length, an autologous nerve graft may not be feasible for extensive nerve injury. Also the harvesting of a nerve graft might lead to morbidity, scarring, sensory loss, and neuroma formation at the donor site [47]. Development of a tissue-engineered nerve graft provides a promising alternative to autologous nerve grafts (Table 3.2). A wide range of synthetic and natural biomaterials are used as scaffolds for bridging nerve gaps and ameliorating nerve regeneration by providing appropriate guidance and neurotrophic support in the regeneration of impaired axons [61–63]. Scaffolds can act as a carrier to deliver SCs for the transplantation into wounded nerve tissue. One study demonstrated a tissue-engineered nerve graft as a promising alternative approach to an autologous nerve graft. In this experiment, to mimic the body's natural regenerative microenvironment, the co-cultured SCs and adipose-derived stem cells (ADSCs) were introduced into a silk fibroin (SF)/collagen scaffold to construct a tissue-engineered nerve conduit (TENC). This TENC scaffold was used to bridge a 1-cm-long sciatic nerve defect in rats. Findings suggest that the effect of treatment with TENCs was similar to that using autologous nerve grafts but superior to that of plain SF/collagen scaffolds [64].

In a similar study, a tissue-engineered nerve graft (TENG) was constructed for the co-culture of dorsal root ganglia and Schwann cells. The SF scaffold seeded with those cells was then implanted to bridge a 1-cm-long sciatic nerve defect in rats for 12 weeks. After 12 weeks of implantation, results showed an improved axonal growth, nerve regeneration, and functional recovery, which was more close to that of the autologous nerve graft than that of the silk fibroin-based scaffold [65]. In another study, silk fibroin scaffolds were improved by functionalization with extracellular molecules. Human placenta-derived collagen and laminin were covalently attached to the SF. The scaffolds with collagen and laminin significantly improved SC adhesion to the SF. This technique improved both cell adhesion and proliferation

3.6 Schwann Cells and Biomaterial for Neural Regeneration

Table 3.2 Schwann cell and biomaterials for neural regeneration

Cell type	Biomaterials	Animal model	Procedure	Results	Refs.
SCs and ADSCs	Tissue-engineered nerve conduit (TENC)	Rat	Co-cultured SCs and adipose-derived stem cells (ADSCs) were introduced into a silk fibroin (SF)/collagen scaffold to construct a TENC. Part of a rat sciatic nerve was removed to create a 1-cm defect. The injured nerve was repaired with TENC, an autograft, or plain SF/collagen scaffold.	Treatment with TENCs was similar to that with autologous nerve grafts but superior to that of plain SF/collagen scaffolds.	[64]
SCs and dorsal root ganglia	Tissue-engineered nerve graft (TENG)	Rat	TENG was constructed using a co-culture system of dorsal root ganglia and SCs, and introduced into a silk-fibroin-based scaffold. TENG was implanted to bridge a 10-mm-long rat sciatic nerve defect.	The implantation of TENG resulted in a functional recovery close to the autologous nerve graft.	[65]
SCs	Polyacrylamide gel	In vitro	Schwann cells were grown on a polyacrylamide gel surface with different stiffnesses.	The 7.45 kPa substrate showed optimal elasticity, which led to improved cell adhesion, survival, proliferation, migration, and production of neurotrophic factors, compared with other substrates.	[108]
SCs	Graphene oxide/polyacrylamide (GO/PAM) composite hydrogels	In vitro	Schwann cells were cultured on GO/PAM composite hydrogels.	The GO/PAM composite hydrogel with 0.4% GO could effectively enhance the attachment and proliferation of Schwann cells and the higher release of biofactors.	[77]

(continued)

Table 3.2 (continued)

Cell type	Biomaterials	Animal model	Procedure	Results	Refs.
SCs	Silk-gold nanocomposite manufactured by adsorbing gold nanoparticles onto silk fibers	Rat	Sciatic nerve was transected and repaired with a conduit by suturing the conduit to both the proximal and distal ends of the nerve stumps.	The conduits promoted adhesion and proliferation of SCs in vitro. The composite did not induce any toxic or immunogenic responses in vivo. The regenerated tissue showed enhanced myelination and improved functional and structural recovery.	[67]
SCs	Polyurethane fabricated from poly(glycerol sebacate)-co-aniline pentamer (PGSAP) polymer	In vitro	Primary SCs were seeded onto the polymer.	Enhanced myelin gene expressions and increased neurotrophin secretion were observed.	[109]

[66]. A silk-gold nanocomposite was manufactured by adsorbing gold nanoparticles (GNPs) onto silk fibers to investigate the function of gold on the biological behavior of SCs in the neural regeneration process. An in vitro study that grew SCs on silk-gold nanocomposite conduits demonstrated that the conduits were able to promote both SC adhesion and SC proliferation. The neural conduits were then implanted in rats with a sciatic nerve injury for more than 18 months. This in vivo study showed that the composite did not induce any toxic or immunogenic responses. Regenerated tissue showed enhanced myelination, improved functional, and structural recovery [67]. A further study based on the above experiment demonstrated that green synthesized GNPs were not cytotoxic to rat SCs [68]. Biodegradable magnesium (Mg) was also shown to induce SCs to secrete nerve growth factor and enhance axon regeneration after compression injury. AZ31 Mg wire (3 mm in diameter and 8 mm in length) was used to bridge both ends of a nerve in a rat model having acute sciatic nerve injury. Results showed an increase in the sciatic functional index, nerve growth factor, p75 neurotrophin, and tyrosine receptor kinases A mRNA expression. This treatment increased the number of nerve fibers on the cross sections [69].

Biomaterial scaffolds are also produced by incorporating different biological molecules, such as proteins and growth factors, to ameliorate their properties [61]. In one experiment, the biocompatibility of a chitosan scaffold was modified using negatively charged heparin and positively charged aminopropyltriethoxysilane (APTES). SCs seeded on the chitosan scaffold modified with a lower heparin concentration showed enhanced attachment capability and developed proliferation [70]. In another in vitro study, a novel scaffold was developed based on bioactive fibers of poly(ε-caprolactone) with the integrated nanotopographical guidance and

NRG1. Purified SCs were cultured in the scaffold for 24 and 72 hours, and results showed directional growth and bipolar differentiation of the cultured SCs [71].

The application of hydrogel in neural repair causes little mechanical stress to the surrounding tissue because its soft and flexible mechanical properties are similar to that of neural tissue. Hydrogel formed by biodegradable and biocompatible materials is highly attractive in the neural engineering research [72]. Hydrogels have been extensively studied in the field of tissue engineering for nerve regeneration [72, 73]. The genipin cross-linked chitosan-sericin hydrogel composite showed high porosity, and adjustable mechanical properties and swelling ratio. Localized delivery of NGF with the chitosan-sericin composite hydrogel up-regulated the mRNA levels of genes like GDNF, EGR2, and the neural cell adhesion molecule (NCAM) in Schwann cells and down-regulated the inflammatory genes like macrophages, tumor necrosis factor alpha (TNFα), and interleukin-1 beta (IL-1β). The regulation of gene expression helped to achieve functional recovery, improved nerve conduction velocity, promoted microstructure restoration, and reduced gastrocnemius muscles dystrophy [74]. A three-dimensional hierarchically aligned fibrin nanofiber hydrogel (AFG) that resembles the architecture and biological function of the native fibrin cable can direct migration and proliferation of SCs and axonal growth. AFG was capable of generating an instructive microenvironment, similar to a native fibrin cable, which promoted the SC cable formation and axonal growth [75]. Wu et al. fabricated the self-assembling peptides by modifying RADA 16-I with two functional motifs, IKAVAV and RGD. The synthesized peptides are called RADA 16-Mix, which could potentially overcome the limitation of RADA 16-I associated with a low pH. The RADA 16-Mix hydrogel was transplanted into the transected rat sciatic nerve gap. Results demonstrated the regeneration of nerves toward the distal position and the induction of more axonal growth resulting in better functional recovery [76]. In one in vitro study, SCs were cultured on a graphene oxide/polyacrylamide (GO/PAM) composite hydrogel. Results showed that GO/PAM in a suitable concentration in PAM hydrogel could effectively promote SC growth [77].

Nanofibers are attractive substrates for nerve repair because they mimic the nerve microenvironment and can provide structural guidance for axonal growth. The advantages of electrospun fibers include the fact that they can have a high surface area-to-volume ratio and controlled packing configurations. Electrospinning techniques have shown potential in developing fibrous scaffolds for nerve repair. Biological molecules can be incorporated into nanofibers to enhance their function in the nerve-repair process. Bhutto et al. fabricated hydrophilic electrospun nanofiber by merging vitamin B5 (50 mg) with an 8% P(LLA-CL)/SF solution. The fabricated nanofiber proved to be relatively elastic and suitable for tissue engineering. SCs seeded in the nanofiber showed enhanced proliferation [78]. Suzuki et al. designed a novel electrospun nanofiber sheet incorporating the active form of vitamin B12 homologs, methylcobalamin (MeCbl). The electrospun nanofiber sheets assisted release in MeCbl for at least 8 weeks when tested in vitro, and no toxic effect on the nanofiber sheets was found in vivo. Transplantation of the nanofiber sheet incorporated with MeCbl in a rat model with sciatic nerve injury promoted the recovery of motor and sensory function, nerve conduction velocity, and myelination [79].

Nune et al. developed a hybrid nanofiber scaffold by combining PLGA and RADA16-I-BMHP. SCs seeded in the peptide-blended nanofiber showed significantly improved cell adhesion, differentiation, and phenotypic expression. The hybrid scaffold also promoted gene expression of neural development markers such as SEM3F, NRP2, and PLX1 [80]. In an experiment conducted by Lv et al., polycaprolactone-electrospun conduits filled with self-assembling peptide RADA16-I and collagen VI protein were used to bridge a 15-mm left sciatic nerve defect in rats. Results showed an increase in the recruitment of macrophages and their polarization, which assist in axonal regeneration and neurological functional recovery [81]. Recently, poly-L-lactide (PLLA) nanofibrous scaffolds coated with multilayers of heparin/collagen encapsulating NGFs via a layer-by-layer (LBL) self-assembling technique were generated. This novel scaffold facilitated sustained release of bioactive NGF from aligned nanofibers, indicating excellent potential in peripheral nerve regeneration [82].

3.7 Role of Schwann Cells in Spinal Cord Injury

Spinal cord injury (SCI) causes the necrosis of neurons and glial cells, and interrupts the neural tracts. Demyelination of axons at the lesion deteriorates functional loss in the SCI. SC myelination is responsible for the axonal myelination in the peripheral nervous system. Schwann cells participate in the remyelination of axons in SCI [83]. Many experiments have shown that following SCI, endogenous SCs migrate and invade the injury site where they associate with regenerating axons [84, 85]. In one study, rat thoracic spinal cord was injured, and the animals survived for up to 6 weeks. It was noted that the injured cavities were filled with nerve fibers and associated SCs, thus indicating SC migration toward the SCI [86]. In another study, after 3 months of cat thoracic SCI, SCs were shown to remyelinate the surviving axons in the dorsal column, thereby showing that SCs migrate and invade the lesion [87]. Remyelination of SCs in the regenerating axons facilitate the formation of the myelin sheath around the newly developed axons, in turn supporting the signal conduction in the regenerated axons [88]. SCs express NGF, BDNF, ciliary neurotrophic factor (CNTF), GDNF, and fibroblast growth factor (FGF). They also express axon growth-supporting cell adhesion molecules and produce axon growth-promoting substrates such as laminin and fibronectin. These factors improve the axon repair process [19].

Schwann cell transplantation into the wounded spinal cord may restore axonal myelination and improve the functional recovery of neural circuitry. SCs are one of the most widely used therapeutic approaches for repairing SCI [19]. SCs can be easily isolated from the peripheral nerve and cultured in vitro. Numerous studies have also shown that SC transplantation can provide a neuroprotective effect and promote axonal regeneration in spinal cord repair [89, 90]. A number of studies have demonstrated the remyelination of grafted SCs for the axons of a wounded spinal cord. To promote injured spinal cord regeneration, SCs were transplanted into the injured neural tissue. Results showed that SC implantation enhanced spinal axonal regeneration [91, 92]. Transplanted SCs can retain their myelination capability in the spinal cord of a xenogenic host. Transplantation of rat SCs into the spinal cords of the myelin of mutant mice resulted in remyelination of spinal axons in the dorsal columns from 2 to 18 weeks after implantation [93]. However, the grafted SCs demonstrated poor migration capacity in the host spinal cord. In one study, the transplantation of SCs into the contused adult rat spinal cord resulted in axonal myelination. However, most grafted SCs did not migrate and remained at the immediate site of injection over time. The poor SC migration ability limited their potential to remyelinate distal axons in the lesion [83, 94].

Although SCs have proven to be effective in SCI regeneration, the effect on facilitating axonal growth is limited. The graft with SCs alone is not able to restore function post-SCI. Additional interventions are required to enhance the axonal regeneration and locomotory function recovery [19, 89]. In one study, SCs extracted from rat sciatic nerve were transplanted into transected rat spinal cord. The rats also received simultaneous injection of edaravone in the vein. The combined treatment of edaravone and SCs showed survival and migration of SCs to the center region of the SCI and demonstrated improved motor and neurophysiological function [95]. Implantation of SCs supplemented with putrescine into rats 1 week after SCI showed enhanced axonal growth and improved locomotion, thereby indicating that polyamine supplement in SCs can be an effective approach for promoting locomotion recovery [96]. In another study, the transplantation of SCs in combination with the anaphylatoxin complement factor 5a (C5a) into rat SCI resulted in improved rear limb functional recovery [97].

The myelination potential of SCs and olfactory ensheathing glia cells (OEGCs) for spinal axons were compared in one previous study. One week after moderate injury of a rat thoracic spinal cord, the SCI was transplanted with either cultured adult rat SCs, OEGCs, or both (SC-OEG). Twelve weeks after SCI and repair, the SC, OEGCs, and SC-OEG groups showed that 51%, 44%, and 44%, respectively, of spinal tissue within a 5-mm-long segment of cord centered at the injury site were spared. The animals treated with SC alone contained more myelinated axons than the other two groups. Thus, the SC graft was more efficient in axonal regeneration [98]. Human telomerase reverse transcriptase gene (hTERT) has also been found to have therapeutic potential in SCI. SCs transfected with a gene vector expressing hTERT were injected into the rat model with acute SCI. Within 1–4 weeks of gene-transfected SC implantation, improved motor function of the lower limb, fewer apoptotic cells, and improved tissue repair were observed [99]. The co-transplantation of SCs and stem cells showed enhanced function of these cells in SCI. In the study, activated SCs and bone marrow mesenchymal stem cells (BMSCs) were obtained from patients having spinal cord injury. The cells were cotransplanted into the epicenter of a rat SCI model. Co-transplantation of the combined cells showed improved functional recovery in the rat's hind limb and reduction in the formation of a glial scar [100].

Biomaterial scaffolds provide a carrier and permissive environment for the transplantation of SCs. In one study, a multichannel scaffold seeded with activated Schwann cells (ASCs) and rat BMSCs were grafted into rats having a 3-mm-wide transection gap in the thoracic spinal cord. Post-transplantation results demonstrated the significant axon regeneration and expression of a mature neuronal marker in rats receiving a multichannel polymer scaffold and those cells [101].

The efficacy of SCs in SCI therapy was evaluated in the Miami Project, clinical trials conducted to examine the potency of transplanting autologous human SCs into the injury epicenter of six patients suffering from subacute SCI. The sural nerve from each patient was harvested and processed to provide the SCs. Cultured autologous SCs from the sural nerve of each patient were injected into the epicenter of spinal lesions. One-year post-transplantation results showed no surgical, medical, or neurological complications, and no adverse events or serious adverse events related to the cell therapy. Also, there was no sign of additional spinal cord damage, mass lesion, or syrinx formation [102, 103].

References

1. Rosso G, Young P, Shahin V. Implications of Schwann cells biomechanics and Mechanosensitivity for peripheral nervous system physiology and pathophysiology. Front Mol Neurosci. 2017;10:345.
2. Jessen KR, Mirsky R. Origin and early development of Schwann cells. Microsc Res Tech. 1998;41(5):393–402.
3. Woodhoo A, Sommer L. Development of the Schwann cell lineage: from the neural crest to the myelinated nerve. Glia. 2008;56(14):1481–90.
4. Gomez-Sanchez JA, et al. Sustained axon–glial signaling induces Schwann cell Hyperproliferation, Remak bundle myelination, and tumorigenesis. J Neurosci. 2009;29(36):11304–15.
5. Wang L, Sanford MT, Xin Z, Lin G, Lue TF. Role of Schwann cells in the regeneration of penile and peripheral nerves. Asian J Androl. 2015;17(5):776–82.
6. Namgung U. The role of Schwann cell-axon interaction in peripheral nerve regeneration. Cells Tissues Organs. 2014;200(1):6–12.
7. Harty BL, Monk KR. Unwrapping the unappreciated: recent progress in Remak Schwann cell biology. Curr Opin Neurobiol. 2017;47:131–7.
8. Stolt CC, Wegner M. Schwann cells and their transcriptional network: evolution of key regulators of peripheral myelination. Brain Res. 2016;1641(Pt A):101–10.
9. Britsch S, et al. The transcription factor Sox10 is a key regulator of peripheral glial development. Genes Dev. 2001;15(1):66–78.
10. Jessen KR, Mirsky R. Schwann cells and their precursors emerge as major regulators of nerve development. Trends Neurosci. 1999;22(9):402–10.
11. Quintes S, et al. Zeb2 is essential for Schwann cell differentiation, myelination and nerve repair. Nat Neurosci. 2016;19(8):1050–9.
12. Wu LM, et al. Zeb2 recruits HDAC-NuRD to inhibit notch and controls Schwann cell differentiation and remyelination. Nat Neurosci. 2016;19(8):1060–72.
13. Busuttil F, Rahim AA, Phillips JB. Combining gene and Stem cell therapy for peripheral nerve tissue engineering. Stem Cells Dev. 2017;26(4):231–8.
14. Hoffman PN. A conditioning lesion induces changes in gene expression and axonal transport that enhance regeneration by increasing the intrinsic growth state of axons. Exp Neurol. 2010;223(1):11–8.
15. Kazakova N, et al. A screen for mutations in zebrafish that affect myelin gene expression in Schwann cells and oligodendrocytes. Dev Biol. 2006;297(1):1–13.

References

16. Maggi SP, Lowe JB 3rd, Mackinnon SE. Pathophysiology of nerve injury. Clin Plast Surg. 2003;30(2):109–26.
17. Waller A. Experiments on the section of the glossopharyngeal and hypoglossal nerves of the frog, and observations of the alterations produced thereby in the structure of their primitive Fibres. Philos Trans R Soc Lond. 1850;140:423–9.
18. van Niekerk EA, Tuszynski MH, Lu P, Dulin JN. Molecular and cellular mechanisms of axonal regeneration after spinal cord injury. Mol Cell Proteom. 2016;15(2):394–408.
19. Oudega M, Xu XM. Schwann cell transplantation for repair of the adult spinal cord. J Neurotrauma. 2006;23(3–4):453–67.
20. You S, Petrov T, Chung PH, Gordon T. The expression of the low affinity nerve growth factor receptor in long-term denervated Schwann cells. Glia. 1997;20(2):87–100.
21. Chen ZL, Yu WM, Strickland S. Peripheral regeneration. Ann Rev Neurosci. 2007;30:209–33.
22. David S, Aguayo AJ. Axonal elongation into peripheral nervous system "bridges" after central nervous system injury in adult rats. Science (New York, NY). 1981;214(4523):931–3.
23. Gaudet AD, Popovich PG, Ramer MS. Wallerian degeneration: gaining perspective on inflammatory events after peripheral nerve injury. J Neuroinflammation. 2011;8:110.
24. Scheib J, Hoke A. Advances in peripheral nerve regeneration. Nat Rev Neurol. 2013;9(12):668–76.
25. Michaelevski I, et al. Signaling to transcription networks in the neuronal retrograde injury response. Sci Signal. 2010;3(130):ra53.
26. Stam FJ, et al. Identification of candidate transcriptional modulators involved in successful regeneration after nerve injury. Eur J Neurosci. 2007;25(12):3629–37.
27. Boyd JG, Gordon T. Neurotrophic factors and their receptors in axonal regeneration and functional recovery after peripheral nerve injury. Mol Neurobiol. 2003;27(3):277–324.
28. Fu SY, Gordon T. The cellular and molecular basis of peripheral nerve regeneration. Mol Neurobiol. 1997;14(1–2):67–116.
29. Jessen KR, Mirsky R. The repair Schwann cell and its function in regenerating nerves. J Physiol. 2016;594(13):3521–31.
30. Gordon T, Tyreman N, Raji MA. The basis for diminished functional recovery after delayed peripheral nerve repair. J Neurosci. 2011;31(14):5325–34.
31. Brushart TM. Nerve repair: Oxford University Press; 2011.
32. He X, et al. A histone deacetylase 3-dependent pathway delimits peripheral myelin growth and functional regeneration. Nat Med. 2018;24(3):338–51.
33. IJ-P J, Jansen K, Gramsbergen A, Meek MF. Transection of peripheral nerves, bridging strategies and effect evaluation. Biomaterials. 2004;25(9):1583–92.
34. Levi AD, et al. The use of autologous Schwann cells to supplement sciatic nerve repair with a large gap: first in human experience. Cell Transplant. 2016;25(7):1395–403.
35. Hendry JM, Alvarez-Veronesi MC, Snyder-Warwick A, Gordon T, Borschel GH. Side-to-side nerve bridges support donor axon regeneration into chronically Denervated nerves and are associated with characteristic changes in Schwann cell phenotype. Neurosurgery. 2015;77(5):803–13.
36. Gordon T, et al. Nerve cross-bridging to enhance nerve regeneration in a rat model of delayed nerve repair. PLoS One. 2015;10(5):e0127397.
37. Hundepool CA, et al. Optimizing decellularization techniques to create a new nerve allograft: an in vitro study using rodent nerve segments. Neurosurg Focus. 2017;42(3):E4.
38. Jiang X, et al. (2016) Effect of frankincense extract on nerve recovery in the rat sciatic nerve damage model. *Evidence-based complementary and alternative medicine : eCAM* 2016:3617216.
39. Ma J, et al. Effect of metformin on Schwann cells under hypoxia condition. Int J Clin Exp Pathol. 2015;8(6):6748–55.
40. Hendry JM, et al. ErbB2 blockade with Herceptin (trastuzumab) enhances peripheral nerve regeneration after repair of acute or chronic peripheral nerve injury. Ann Neurol. 2016;80(1):112–26.

41. Ju DT, et al. Nerve regeneration potential of Protocatechuic acid in RSC96 Schwann cells by induction of cellular proliferation and migration through IGF-IR-PI3K-Akt signaling. Chin J Physiol. 2015;58(6):412–9.
42. Liu H, et al. Salidroside promotes peripheral nerve regeneration based on tissue engineering strategy using Schwann cells and PLGA: in vitro and in vivo. Sci Rep. 2017;7:39869.
43. Tuffaha SH, et al. Therapeutic augmentation of the growth hormone axis to improve outcomes following peripheral nerve injury. Expert Opin Ther Targets. 2016;20(10):1259–65.
44. Saceda J, et al. Effect of recombinant human growth hormone on peripheral nerve regeneration: experimental work on the ulnar nerve of the rat. Neurosci Lett. 2011;504(2):146–50.
45. Devesa P, et al. Growth hormone treatment enhances the functional recovery of sciatic nerves after transection and repair. Muscle Nerve. 2012;45(3):385–92.
46. Tuffaha SH, et al. Growth hormone therapy accelerates axonal regeneration, promotes motor Reinnervation, and reduces muscle atrophy following peripheral nerve injury. Plast Reconstr Surg. 2016;137(6):1771–80.
47. Muheremu A, Ao Q. Past, present, and future of nerve conduits in the treatment of peripheral nerve injury. Biomed Res Int. 2015;2015:237507.
48. Yi S, et al. Regulation of Schwann cell proliferation and migration by miR-1 targeting brain-derived neurotrophic factor after peripheral nerve injury. Sci Rep. 2016;6:29121.
49. Lin G, et al. Brain-derived neurotrophic factor promotes nerve regeneration by activating the JAK/STAT pathway in Schwann cells. Transl Androl Urol. 2016;5(2):167–75.
50. de Winter F, et al. Gene therapy approaches to enhance regeneration of the injured peripheral nerve. Eur J Pharmacol. 2013;719(1–3):145–52.
51. Zhang Z, Wang H, Sun Y, Li H, Wang N. Klf7 modulates the differentiation and proliferation of chicken preadipocyte. Acta Biochim Biophys Sin. 2013;45(4):280–8.
52. Zou H, Ho C, Wong K, Tessier-Lavigne M. Axotomy-induced Smad1 activation promotes axonal growth in adult sensory neurons. J Neurosci. 2009;29(22):7116–23.
53. Wang Y, et al. Sciatic nerve regeneration in KLF7-transfected acellular nerve allografts. Neurol Res. 2016;38(3):242–54.
54. Wang Y, et al. KLF7-transfected Schwann cell graft transplantation promotes sciatic nerve regeneration. Neuroscience. 2017;340:319–32.
55. Marquardt LM, et al. Finely tuned temporal and spatial delivery of GDNF promotes enhanced nerve regeneration in a long nerve defect model. Tissue Eng Part A. 2015;21(23–24):2852–64.
56. Ziegler L, Grigoryan S, Yang IH, Thakor NV, Goldstein RS. Efficient generation of schwann cells from human embryonic stem cell-derived neurospheres. Stem Cell Rev. 2011;7(2):394–403.
57. Cui L, et al. Transplantation of embryonic stem cells improves nerve repair and functional recovery after severe sciatic nerve axotomy in rats. Stem Cells (Dayton, Ohio). 2008;26(5):1356–65.
58. Lee EJ, et al. Regeneration of peripheral nerves by transplanted sphere of human mesenchymal stem cells derived from embryonic stem cells. Biomaterials. 2012;33(29):7039–46.
59. Wang H, Wu J, Zhang X, Ding L, Zeng Q. Study of synergistic role of allogenic skin-derived precursor differentiated Schwann cells and heregulin-1beta in nerve regeneration with an acellular nerve allograft. Neurochem Int. 2016;97:146–53.
60. Gu Y, et al. Basic fibroblast growth factor (bFGF) facilitates differentiation of adult dorsal root ganglia-derived neural stem cells toward Schwann cells by binding to FGFR-1 through MAPK/ERK activation. J Mol Neurosci. 2014;52(4):538–51.
61. Gu X, Ding F, Yang Y, Liu J. Construction of tissue engineered nerve grafts and their application in peripheral nerve regeneration. Prog Neurobiol. 2011;93(2):204–30.
62. Evans GR. Peripheral nerve injury: a review and approach to tissue engineered constructs. Anat Rec. 2001;263(4):396–404.
63. Johnson EO, Soucacos PN. Nerve repair: experimental and clinical evaluation of biodegradable artificial nerve guides. Injury. 2008;39(Suppl 3):S30–6.

64. Xu Y, et al. A silk fibroin/collagen nerve scaffold seeded with a co-culture of Schwann cells and adipose-derived stem cells for sciatic nerve regeneration. PLoS One. 2016;11(1): e0147184.
65. Tang X, et al. Bridging peripheral nerve defects with a tissue engineered nerve graft composed of an in vitro cultured nerve equivalent and a silk fibroin-based scaffold. Biomaterials. 2012;33(15):3860–7.
66. Schuh CM, Monforte X, Hackethal J, Redl H, Teuschl AH. Covalent binding of placental derived proteins to silk fibroin improves schwann cell adhesion and proliferation. J Mater Sci Mater Med. 2016;27(12):188.
67. Das S, et al. In vivo studies of silk based gold nano-composite conduits for functional peripheral nerve regeneration. Biomaterials. 2015;62:66–75.
68. Das S, et al. Data in support of in vivo studies of silk based gold nano-composite conduits for functional peripheral nerve regeneration. Data Brief. 2015;4:315–21.
69. Li BH, Yang K, Wang X. Biodegradable magnesium wire promotes regeneration of compressed sciatic nerves. Neural Regen Res. 2016;11(12):2012–7.
70. Li G, Zhang L, Yang Y. Tailoring of chitosan scaffolds with heparin and gamma-aminopropyltriethoxysilane for promoting peripheral nerve regeneration. Colloids Surf B Biointerfaces. 2015;134:413–22.
71. Tonazzini I, Moffa M, Pisignano D, Cecchini M. Neuregulin 1 functionalization of organic fibers for Schwann cell guidance. Nanotechnology. 2017;28(15):155303.
72. Mandal BB, Kapoor S, Kundu SC. Silk fibroin/polyacrylamide semi-interpenetrating network hydrogels for controlled drug release. Biomaterials. 2009;30(14):2826–36.
73. Jing S, Jiang D, Wen S, Wang J, Yang C. Preparation and characterization of collagen/silica composite scaffolds for peripheral nerve regeneration. J Porous Mater. 2014;21(5):699–708.
74. Zhang L, et al. Sustained local release of NGF from a chitosan-Sericin composite scaffold for treating chronic nerve compression. ACS Appl Mater Interfaces. 2017;9(4):3432–44.
75. Du J, et al. Prompt peripheral nerve regeneration induced by a hierarchically aligned fibrin nanofiber hydrogel. Acta Biomater. 2017;55:296.
76. Wu X, et al. Functional self-assembling peptide nanofiber hydrogel for peripheral nerve regeneration. Regen Biomater. 2017;4(1):21–30.
77. Li G, et al. Preparation of graphene oxide/polyacrylamide composite hydrogel and its effect on Schwann cells attachment and proliferation. Colloids Surf B Biointerfaces. 2016;143:547–56.
78. Bhutto MA, et al. Fabrication and characterization of vitamin B5 loaded poly (l-lactide-co-caprolactone)/silk fiber aligned electrospun nanofibers for schwann cell proliferation. Colloids Surf B Biointerfaces. 2016;144:108–17.
79. Suzuki K, et al. Electrospun nanofiber sheets incorporating methylcobalamin promote nerve regeneration and functional recovery in a rat sciatic nerve crush injury model. Acta Biomater. 2017;53:250–9.
80. Nune M, Krishnan UM, Sethuraman S. PLGA nanofibers blended with designer self-assembling peptides for peripheral neural regeneration. Mater Sci Eng C Mater Biol Appl. 2016;62:329–37.
81. Lv D, Zhou L, Zheng X, & Hu Y (2017) Sustained release of collagen VI potentiates sciatic nerve regeneration by modulating macrophage phenotype. *The European journal of neuroscience*.
82. Zhang K, Huang D, Yan Z, Wang C. Heparin/collagen encapsulating nerve growth factor multilayers coated aligned PLLA nanofibrous scaffolds for nerve tissue engineering. J Biomed Mater Res A. 2017;105:1900.
83. Guest JD, Hiester ED, Bunge RP. Demyelination and Schwann cell responses adjacent to injury epicenter cavities following chronic human spinal cord injury. Exp Neurol. 2005;192(2):384–93.
84. Wang ZH, Walter GF, Gerhard L. The expression of nerve growth factor receptor on Schwann cells and the effect of these cells on the regeneration of axons in traumatically injured human spinal cord. Acta Neuropathol. 1996;91(2):180–4.

85. Brook GA, et al. Spontaneous longitudinally orientated axonal regeneration is associated with the Schwann cell framework within the lesion site following spinal cord compression injury of the rat. J Neurosci Res. 1998;53(1):51–65.
86. Beattie MS, et al. Endogenous repair after spinal cord contusion injuries in the rat. Exp Neurol. 1997;148(2):453–63.
87. Blight AR, Young W. Central axons in injured cat spinal cord recover electrophysiological function following remyelination by Schwann cells. J Neurol Sci. 1989;91(1–2):15–34.
88. Pinzon A, Calancie B, Oudega M, Noga BR. Conduction of impulses by axons regenerated in a Schwann cell graft in the transected adult rat thoracic spinal cord. J Neurosci Res. 2001;64(5):533–41.
89. Kanno H, Pearse DD, Ozawa H, Itoi E, Bunge MB. Schwann cell transplantation for spinal cord injury repair: its significant therapeutic potential and prospectus. Rev Neurosci. 2015;26(2):121–8.
90. Young W. Spinal cord regeneration. Cell Transplant. 2014;23(4–5):573–611.
91. Liu G, Wang X, Shao G, Liu Q. Genetically modified Schwann cells producing glial cell line-derived neurotrophic factor inhibit neuronal apoptosis in rat spinal cord injury. Mol Med Rep. 2014;9(4):1305–12.
92. Iannotti C, et al. Glial cell line-derived neurotrophic factor-enriched bridging transplants promote propriospinal axonal regeneration and enhance myelination after spinal cord injury. Exp Neurol. 2003;183(2):379–93.
93. Duncan ID, Aguayo AJ, Bunge RP, Wood PM. Transplantation of rat Schwann cells grown in tissue culture into the mouse spinal cord. J Neurol Sci. 1981;49(2):241–52.
94. Totoiu MO, Keirstead HS. Spinal cord injury is accompanied by chronic progressive demyelination. J Comp Neurol. 2005;486(4):373–83.
95. Zhang S-Q, et al. Edaravone combined with Schwann cell transplantation may repair spinal cord injury in rats. Neural Regen Res. 2015;10(2):230–6.
96. Iorgulescu JB, et al. Acute Putrescine supplementation with Schwann cell implantation improves sensory and serotonergic axon growth and functional recovery in spinal cord injured rats. Neural Plast. 2015;2015:186385.
97. Zhang SQ, et al. Improvements in neuroelectrophysiological and rear limb functions in rats with spinal cord injury after Schwann cell transplantation in combination with a C5a receptor antagonist. Gene Mol Res. 2015;14(4):15158–68.
98. Takami T, et al. Schwann cell but not olfactory ensheathing glia transplants improve hindlimb locomotor performance in the moderately contused adult rat thoracic spinal cord. J Neurosci. 2002;22(15):6670–81.
99. Zhang SQ, et al. Transplantation of human telomerase reverse transcriptase gene-transfected Schwann cells for repairing spinal cord injury. Neural Regen Res. 2015;10(12):2040–7.
100. Ban DX, et al. Combination of activated Schwann cells with bone mesenchymal stem cells: the best cell strategy for repair after spinal cord injury in rats. Regen Med. 2011;6(6):707–20.
101. Yang E-Z, et al. Multichannel polymer scaffold seeded with activated Schwann cells and bone mesenchymal stem cells improves axonal regeneration and functional recovery after rat spinal cord injury. Acta Pharmacol Sin. 2017;38(5):623–37.
102. Anderson KD, et al. Safety of autologous human Schwann cell transplantation in subacute thoracic spinal cord injury. J Neurotrauma. 2017;34(21):2950–63.
103. Xu XM. Breaking news in spinal cord injury research: FDA approved phase I clinical trial of human, autologous schwann cell transplantation in patients with spinal cord injuries. Neural Regen Res. 2012;7(22):1685–7.
104. Shi H, et al. Derivation of Schwann cell precursors from neural crest cells resident in bone marrow for cell therapy to improve peripheral nerve regeneration. Biomaterials. 2016;89:25–37.
105. Latasa MJ, Jimenez-Lara AM, Cosgaya JM. Retinoic acid regulates Schwann cell migration via NEDD9 induction by transcriptional and post-translational mechanisms. Biochim Biophys Acta. 2016;1863(7 Pt A):1510–8.

References

106. Wu W, Liu Y, Wang Y. Sam68 promotes Schwann cell proliferation by enhancing the PI3K/Akt pathway and acts on regeneration after sciatic nerve crush. Biochem Biophys Res Commun. 2016;473(4):1045–51.
107. Yi S, et al. miR-sc3, a novel microRNA, promotes Schwann cell proliferation and migration by targeting Astn1. Cell Transplant. 2016;25(5):973–82.
108. Gu Y, et al. The influence of substrate stiffness on the behavior and functions of Schwann cells in culture. Biomaterials. 2012;33(28):6672–81.
109. Wu Y, Wang L, Guo B, Shao Y, Ma PX. Electroactive biodegradable polyurethane significantly enhanced Schwann cells myelin gene expression and neurotrophin secretion for peripheral nerve tissue engineering. Biomaterials. 2016;87:18–31.

Chapter 4
Stem Cell- and Biomaterial-Based Neural Repair for Enhancing Spinal Axonal Regeneration

Pranita Kaphle, Li Yao, and Joshua Kehler

Abstract Axonal damage in spinal cord injury (SCI) results in functional impairment and neurological disorders. Stem cell transplantation has been studied for the therapy of SCI because of their proliferation and differentiation capacity. Stem cell transplantation is potentially able to replace the neuron and glia loss in damaged neural tissue to restore the functional connection of neural tract. The extracellular matrix and several neurotrophic factors produced by stem cells can improve axonal growth. The stem cells-derived oligodendrocytes promote myelin formation for the remaining and newly grown neural axons. Biomaterials have offered an effective carrier means for stem cell transplantation following SCI. Various complications and pathophysiological concerns that arise during axonal regeneration following SCI may be overcome by the co-transplantation of biomaterial scaffolds fabricated from natural and synthetic polymers. Biomaterial scaffolds may effectively ameliorate cell death following transplantation and preventing scar formation. In this chapter, we review the advance in the research of stem cell transplantation including embryonic stem cells, neural stem cells, induced pluripotent stem cells, and mesenchymal stem cells in neural repair and regeneration.

Keywords Electric field · Guided migration · Axonal growth · Polarity · Cell division · Spinal cord injury · Neurogenesis · Stem cell · Signaling pathway · RNA sequencing

4.1 Stem Cell Therapy for Axonal Regeneration in Spinal Cord Repair

Axonal injury results in functional impairment and neurological disorders. Generally, complete spinal cord injury (SCI) refers to the loss of nervous tissue and consequently the loss of sensory and motor functions and the retention of some neuron function indicates incomplete injury. The most frequent type of traumatic SCI is acute compression of the spinal cord [1]. In SCI, there is less chance for the survival of a neuronal cell body once a neurite detaches from it. After injury, the axonal segment separates where the distal end of the axons disconnect from the neuronal cell bodies by a process called Wallerian, or orthograde, degeneration [2]. The nervous system maintains a limited capacity of regeneration. In the peripheral nervous system (PNS), Schwann cells (SCs) promote a suitable environment to regenerate axons after nerve injury. The distal end of the axon degenerates, and SCs cease biosynthesis of new myelin membrane [3]. The fragments produced by degradation of myelin sheaths after degeneration of an axon inhibit axonal growth [4]. However, some stimulatory factors allow axon regeneration in the PNS after injury. At the injury site, cytokines secreted by SCs attract macrophages and neutrophils to remove debris and thereby create a suitable environment for axons to regenerate [2]. In contrast, the accumulation of astrocytes in response to spinal cord injury creates a barrier to axon regeneration in the central nervous system (CNS) [5–7]. After SCI, the keratin sulfate and chondroitin sulfate proteoglycans produced by reactive cells joining the extracellular matrix after the injury block axonal growth in CNS [8, 9]. However, the transplantation of a peripheral nerve graft in the adult CNS allows axons to re-grow. Studies have been undertaken to study the mechanism of PNS regeneration, and the knowledge gained has been used in the CNS for treating neurodegenerative diseases [10]. Furthermore, the application of different types of stem cells, including embryonic stem cells (ESCs), induced pluripotent stem cells (iPSCs), neural stem cells (NSCs), and mesenchymal stem cells (MSCs), to promote axon regeneration in the CNS has been investigated. Stem cell-based therapy has been initiated with the hope of effective treatment of spinal cord injury.

Stem cells possess the potential for self-renewal, continuously dividing asymmetrically. Because of their proliferation and differentiation capacity, stem cell transplantation has been studied for the treatment of SCI. After transplantation, stem cells are potentially able to differentiate into neurons and glia to replace damaged neural tissue [11]. The transplanted stem cells may produce an extracellular matrix and several neurotrophic factors. The interaction of transplanted stem cells with surrounding tissue changes the microenvironment at the injury site and facilitates axonal regeneration [12, 13]. The transplanted stem cell differentiates into oligodendrocytes and gliocytes, and promotes myelin formation around the remaining and newly grown neural axons [14–16].

4.1.1 Embryonic Stem Cells

Embryonic stem cells, which are pluripotent cells, are derived from the inner cell mass of a developing blastocyst embryo undergoing indefinite replication. ESCs can be differentiated into various cell types. It was reported that rodent embryonic stem cells can generate three different types of embryonic germ layers [17, 18] and differentiate into different cell types in vitro. ESCs have been investigated in the treatment of spinal cord injury [19]. Transplantation of ESC-derived oligodendrocyte progenitor cells (OPCs) to the site of adult rat spinal cord injury enhances the remyelination of spared axons and promotes the improvement of motor function

[20, 21]. The remyelination of axons contributes to restoring the function of action-potential conduction [22] and locomotor function [23, 24]. One study demonstrated that the transplantation of human ESC-derived OPCs into an adult rat spinal cord injury site 7 days following the injury enhanced the process of remyelination and promoted locomotor recovery. However, transplantation of the cells at 10 months post-SCI did not enhance remyelination or locomotor recovery [20]. The graft of ESC-derived motoneurons facilitates axon extension along the entire length of the implanted ventral root and promotes reinnervation of the muscles while restoring limb locomotion function [25].

ESCs transfected with the neural cell adhesion molecule L1 gene enhanced the survival of grafted cells within the lesion site of mouse spinal cord with compression injury as it promoted axonal growth [26]. The transplantation of ESC-derived neural aggregates, primarily neurons and radial glial cells over expressing neural cell adhesion molecule L1, significantly improved motor functions and increased cell survival [27]. Human embryonic neural progenitors were engineered to express neurogenin 2, a transcription factor that plays a key role in CNS development and the specification of neuronal differentiation [28], and they were transplanted into a spinal cord lesion of a rat model. The graft of the cells improved spinal motor recovery and axonal regeneration [29]. Human embryonic stem cell (hESC)-derived neural stem cells, when transplanted with a collagen scaffold into the hemisection of a rat model, differentiated successfully into neurons and glial cells in vivo and promoted locomotor recovery with prolonged graft survival [30]. The combinatorial treatment of hESC-derived neural progenitor cells and Schwann cells in a rat SCI model significantly improved motor function improvement. The transplantation of hESC-derived motoneuron progenitor cells [31] and oligodendrocyte progenitor cells [20, 21, 32] in SCI rat models showed neuronal regeneration and reduced acute inflammation. The transplantation of hESC-derived motoneuron progenitor cells, OPCs [33], and olfactory ensheathing cells [34] resulted in effective axonal remyelination and functional recovery. The co-transplantation of hESC-derived neural progenitor cells and SCs provide synergistic effects and result in functional recovery by enhancing neuronal differentiation and suppressing glial differentiation (Table 4.1).

The remyelination of the spared axons plays a significant role in regeneration following injury. Transplantation of ESC-derived neurospheres increased the functional recovery in the mouse model after SCI. This study begins with the generation of two types of neurospheres: primary and passaged secondary. The neurospheres were transplanted into the spinal cord-injured rat model 9 days after injury, subacutely. The remyelination of axons, angiogenesis, and functional recovery was observed in the passaged secondary neurospheres group, yet was absent in the primary neurospheres group. This may be due to the secretion of neurotrophic growth factors from the gliogenic neurospheres, which promote the recovery from SCI [57].

Although remyelination enhanced axonal regeneration, the addition of neurotrophic factors has been used to enhance functional recovery and promote axonal regeneration in the disrupted pathways. The transplantation of pre-differentiated human mesenchymal stem cell (hMSC)-derived oligodendrocytes initiates activation of brain-derived neurotrophic factor (BDNF) and interleukin-6 (IL-6) signaling

Table 4.1 Transplantation of stem cells for repair of injured spinal cord

Cell type	Model	Procedure	Results	Refs.
hESC	Rat	Transplantation of hESC-derived OPCs into injured adult rat spinal cord.	Transplanted cell survival, improved locomotor ability, and remyelination.	[20]
hESC	Rat	Transplantation of hESC-derived OPCs into rats for repair of sensory tracts after contusive SCI.	Functional recovery of sensory pathways in spinal cord.	[35]
hESC	rat	Transplantation of hESC-OPCs into rat spinal cord with acute cervical SCI.	Improved histological outcomes correlating with improved recovery.	[21]
hESC	Rat	Differentiation of hESCs into high-purity OPCs, and transplantation of OPCs into spinal cord injury sites in adult rats.	Remyelination and functional repair at injury sites.	[36]
ESC and BMSC	Mouse	Transplantation of ESCs with BMSCs to prevent tumor formation in SCI model.	Undifferentiated ESCs induced by BMSCs to differentiate into neuronal lineage by neurotrophic factor production, resulting in suppression of tumor formation.	[37]
ESC	Rat	Transplantation of mouse ESCs after neural differentiation into rat spinal cord 9 days after traumatic injury.	Survival of grafted cells and differentiation into astrocytes, oligodendrocytes, and neurons after 2–5 weeks, and migration as far as 8 mm away from lesion edge.	[38]
ESC	Mouse	Transplantation of pre-differentiated ESCs into injured mouse spinal cord 1 week after surgery.	Activation of both BDNF and IL-6 signaling pathways in host tissue, leading to activation of cAMP/PKA, phosphorylation of cofilin and synapsin I, and promotion of regenerative growth and neuronal survival.	[39]
hESC	Rat	Grafting of hESC-derived OPC and/or motoneuron progenitors into site of injured spinal cord in acute phase.	Functional recovery and survival of injured spinal cord of adult rats after complete transaction.	[33]
hESC	Rat	Injection of hESC-derived OPCs into rat spinal cord 3 and 24 h following SCI to study the survival and migration route toward areas of injury.	Survival of ESC-derived OPCs injected at the center of injury and migration away from injection sites one week post-surgery.	[32]
hESC	Rat	Implantation of xenografted hESC-NPC seeded in collagen scaffolds in adult rats subjected to hemisection SCI.	Improvement in recovery of hindlimb locomotor function and sensory responses in adult rat model of SCI.	[30]

(continued)

4.1 Stem Cell Therapy for Axonal Regeneration in Spinal Cord Repair

Table 4.1 (continued)

Cell type	Model	Procedure	Results	Refs.
NSC	Rat	Transduction of NSCs with neurogenin-2 before transplantation, and then cell grafting into injured rat thoracic spinal cord.	Increased amounts of myelin at injured area, and recovery of hindlimb locomotor function and hindlimb sensory responses.	[40]
NSC	Rat	Injection of four groups of rat spinal cord transected at T9 level with DMEM/F12 solution, NSCs, OECs, and NSCs + OECs 7 days post-SCI.	Significant improvement of hindlimb locomotor function of rats in co-transplantation group, compared with that in other three groups, and OEC-promoted NSC differentiation into neurons.	[41]
NSC	Rat/ Mice	Transplantation of NSPCs derived from adult spinal cord of transgenic rats into two models of demyelination: focal and congenital.	Differentiation of NSPCs into oligodendrocytes or Schwann-like cells, and axon myelination of both cell types in demyelinated and dysmyelinated adult spinal cord.	[42]
NSC	Rat	Genetic modification of hNSCs to express retroviral vector encoding Olig2 transcription factor, and then transplantation into rat contusive spinal cord injury model.	Increase in volume of spared white matter, reduction in cavity volume, and increase in thickness of myelin sheath around axons in spared white matter.	[43]
NSC	Rat	Transplantation of hNSCs into spinal cord with T10 contusion injury model after pre-differentiation of hNSCs into cholinergic neurons, either on same day, or 3 or 9 days after injury.	Tendency of primed hNSCs to survive better and differentiate at higher rate into neurons and oligodendrocytes than unprimed control cells.	[44]
NSC	Rat	Grafting of NSCs from human fetal spinal cord into normal or injured lumbar cord of adult nude rats, and observation of large-scale differentiation of these cells into neurons.	Formation of axons and synapses, establishment of extensive contacts with host motor neurons, and substantial neuronal differentiation in normal and injured adult spinal cord.	[45]
NSC	Rat	Isolation of NSCs from embryonic day-14 spinal cord, and embedding into fibrin-containing growth factors for transplantation in lesion cavity of injured spinal cord.	Differentiation of NSCs grafted in lesion cavity into glial cells and neurons, which extended axons into host spinal cord over long distances.	[46]

(continued)

Table 4.1 (continued)

Cell type	Model	Procedure	Results	Refs.
NSC	Rat	Specific functionalization of sodium hyaluronate collagen scaffold with nutrient-bFGF, and transplantation of both bFGF-CRS and NSCs into CA1 zone of traumatic brain-injured area of adult rats.	Observation that bFGF-CRS provided an optimal microenvironment for promoting survival, neuronal differentiation of transplanted NSCs, and functional synapse formation among iN cells, and between iN cells and the host brain tissue in TBI rats, consequently leading to recovery of their cognitive function.	[47]
NSC	Rat	Self-assembly of RADA16-IKVAV into hydrogel, whereby the extended IKVAV sequence can serve as a signal or guiding cue to direct encapsulated NSCs. Also, injection of peptide solution into rat brain.	Neuronal differentiation of encapsulating NSCs, and improvement in brain tissue regeneration after 6 weeks post-transplantation.	[48]
hiPSC	Rat	Differentiation of hiPSCs into NSCs and grafting into adult immunodeficient rats after spinal cord injury.	Survival of iPSCs after 3 months and their differentiation into neurons and glia, and the extension of thousands of axons from the lesion site over virtually the entire length of the rat CNS. Extension of iPSC-derived axons through adult white matter of the injured spinal cord, frequently penetrating gray matter and forming synapses with rat neurons.	[16]
iPSC	Mouse	Transplantation of iPSC-derived neurospheres, which had been pre-evaluated as "safe" nontumorigenic or "unsafe" cells into spinal cord after contusive injury.	Differentiation of safe iPSC-derived neurospheres into all three neural lineages without forming teratomas or other tumors. Promotion of locomotor function recovery by remyelination and induced axonal regrowth. Evidence of "unsafe" cells showing robust teratoma formation and sudden locomotor functional loss after functional recovery in SCI model.	[49]
hiPSC	Mouse	Establishment of hiPSC clone from adult human dermal fibroblasts by retroviral transduction of four reprogramming factors—Oct3/4, Sox2, Klf4, and c-Myc—and then transplantation into non-obese diabetic-severe combined immunodeficient (NOD-SCID) SCI model mice.	Survival, migration, and differentiation of grafted hiPSC-NSs into three neural lineages in injured spinal cord, and also promotion of angiogenesis and axonal re-growth. Preservation of myelination, some formation of synapses with host mouse neurons, and also promotion of functional recovery without tumor formation.	[50]

(continued)

4.1 Stem Cell Therapy for Axonal Regeneration in Spinal Cord Repair

Table 4.1 (continued)

Cell type	Model	Procedure	Results	Refs.
hiPSC	Mouse	Transplantation of hiPS-lt-NES (long-term self-renewing neuroepithelial-like stem cells) cells into injured spinal cord. Also, transplantation of hiPS-lt-NES cells, which express luciferase and GFP, into injured spinal cord to trace survival of transplanted cells.	Improvement in functional recovery of hindlimbs. Survival and differentiation of transplanted hiPS-lt-NES cells in injured spinal cord of NOD-SCID mice.	[51]
iPSC	Mouse	Differentiation of iPSC-derived astrocytes using neural stem sphere (NSS) method and injection three and 7 days after rat spinal contusion-based SCI. Also, injection of control rats with DMEM in same manner.	Survival of transplant recipients for 8 weeks without tumor formation. Stretching of transplanted-cell processes along the longitudinal axis, but non-merging with processes of host GFAP-positive astrocytes.	[52]
MSC	Rat	Implantation of three rat groups with either untreated human umbilical mesenchymal stem cells (HUMSCs) or HUMSCs treated with neuronal conditioned medium (NCM) for either three or 6 days (referred to as NCM-3 days and NCM-6 days, respectively). Also, non-implantation of control group with HUMSCs in transected spinal cord.	Significant improvement in locomotion 3 weeks after transplantation in all three groups receiving HUMSCs (stem cell, NCM-3 days, and NCM-6 days groups). Survival of transplanted HUMSCs for 16 weeks, and production of large amounts of human neutrophil-activating protein-2, neurotrophin-3, basic fibroblast growth factor, glucocorticoid-induced tumor necrosis factor receptor, and vascular endothelial growth factor receptor 3 in host spinal cord, which may help spinal cord repair.	[53]
MSC	Rat	Intravenous administration of rat MSCs derived from bone marrow at various time points after induction of severe contusive SCI, systematic delivery at varied time points (6 h to 28 days) after SCI.	Expression of neural or glial markers by limited number of cells derived from MSCs at injury site. Greater locomotor recovery improvement in MSC-treated groups than controls.	[54]
MSC	Rat	Co-transplantation of ASCs and BMSCs into epicenter of injured rat spinal cord.	Promotion of functional recovery in rat hindlimbs, and reduction in formation of glial scar, and remyelination of injured axons.	[55]

(continued)

Table 4.1 (continued)

Cell type	Model	Procedure	Results	Refs.
MSC	Rat	Subjection of rats to incomplete SCI with contusion at T9 level. Also, transplantation of MSCs or neurally differentiated MSCs (NMSCs) derived from bone marrow into injured spinal cord.	Effective functional recovery of NMSC-treated rats after SCI. Clear tendency of motor recovery after transplantation of MSCs. Significant improvement of NMSC-treated rats in Basso–Beattie–Bresnahan locomotor rating scores (BBB) scores, and significantly shortened initial latency, N1 latency, and P1 latency of SSEPs compared to control group.	[56]

pathways. Studies demonstrate that IL-6 plays an important role in cell–cell signaling within the CNS. In addition, it protects the neurons from insult, coordinates the neuro-immune response, and facilitates neuronal differentiation and growth [58]. During neural development, BDNF and its receptor are expressed in the developing nervous system as well as in the adult nervous system. Following SCI, the expression of BDNF decreases dramatically, taking 3–4 weeks to restore its levels. However, the transplantation of pre-differentiated ESCs promotes the expression of BDNF and IL-6 signaling pathways that lead to the activation of cAMP/PKA. The cAMP/PKA then inactivates RhoA, which in turn inactivates LIMK1/2, inhibiting cofilin phosphorylation and thereby facilitating neuronal regeneration [39]. hESC-derived oligodendrocyte progenitor cells secrete various types of functional molecules including midkine, hepatocyte growth factor (HGF), activin A, transforming growth factor-beta2 (TGF-beta2), and BDNF, all of which assist in the process of neural regeneration [59].

4.1.2 Neural Stem Cells

Neural stem cells are multipotent cells committed to the neural lineage that can perform self-renewal and be readily expanded in vitro. They were first isolated from the striated tissue and subventricular zone of the mouse brain [60, 61]. NSCs are found in both the fetal and adult CNS [62]. The isolation of adult NSCs in mammals was first reported in 1992 by Reynolds and Weiss [60]. NSCs can differentiate into neurons, oligodendrocytes, and astrocytes, both in vivo and in vitro [63–65]. The ependymal cells lining the central canal act as NSCs. In vivo, SCI induces the proliferation of rarely dividing ependymal cells 50 times greater after the first day of injury [66]. Ependymal cells then leave the canal region, migrate toward the injury site within 3 days, and contribute to glial scar formation with the down-regulation of different ependymal-derived cell markers [67]. Furthermore, ependymal cell progeny differentiates into astrocytes and myelinating oligodendrocytes. The transplantation of ependymal stem progenitor cells derived from the injured spinal cord demonstrated a functional motor recovery post-SCI [68]. NSCs can differentiate into oligodendrocytes both in vivo [42, 69] and in vitro [70]. The differentiation of NSCs into OPCs promotes motor and sensory recovery with an increase in axonal remyelination [71]. The transplanted NSCs differentiate mostly into oligodendrocytes and astrocytes, and contribute to the remyelination of axons and recovery of the SCI [72]. Although NSCs have the potential of differentiation into neurons, oligodendrocytes, and astrocytes in vitro [73], it is believed that, following SCI, endogenous NSCs differentiate mostly into oligodendrocytes and astrocytes [72]. Oligodendrocytes are able to remyelinate the axons in white matter, and astrocytes may secrete many neurotrophic factors supporting axonal regeneration and cell survival [64, 74, 75]. The co-transplantation of NPCs and olfactory ensheathing cells (OECs) further promote functional recovery [41]. The graft of mouse brain-derived NPCs into injured rat spinal cord 2 weeks-post injury led to the generation of mature oligodendrocytes that myelinated the injured axons. The transplanted NSCs survived, migrated, and promoted some functional recovery. In contrast, NSCs transplanted 8 weeks after injury did not survive [76]. The chronic progressive demyelination because of delayed loss of oligodendrocytes complicates contusive spinal cord injury. Therefore, NSCs were genetically modified in order to enhance the extent of myelination. The NSCs were transfected with Olig2 transcription factor with the help of a retrovirus and then transplanted into the contusive spinal cord. The grafted NSCs enhanced locomotory recovery with an increase in myelination and reduction in lesion cavity [43]. The remyelination and synaptic contact reformation of the grafted NSCs are equally important to restore the spinal cord circuitry and functional recovery. NSCs from human fetal spinal cord that were grafted into the lumbar cord of adult nude rats resulted in extensive synaptic contacts with host motor neurons and differentiation into neuron and axon regeneration. This outcome might be due to the restoration of neural circuitry in the injured spinal cord [45].

Transplantation of human fetal neural stem progenitor cells (NSPCs) into the adult rat spinal cord showed their extensive differentiation into neurons [45]. The differentiation fate of hNSCs in vivo depends on the transplantation timepoint after injury and might be influenced by the pre-differentiation treatment before transplantation. In addition, differentiation fate of NSCs is determined by the spinal cord microenvironment. NSCs differentiate into neurons if they are centrally located or into astrocytes if they are located under the pia membrane. Furthermore, a lesion in the white matter of the spinal cord promotes NSCs to differentiate into astrocytes [45]. Although the transplanted NSCs showed survival and differentiation into astrocytes, direct transplantation of NSCs is not always efficient in the recovery of spinal cord injury. The transplantation of fetal NPCs into a rat lesion site showed limited recovery, probably due to lack of axon regeneration and low neuronal differentiation beyond the lesion site. Furthermore, transplantation of human fetal brain NSPCs expressing Galectin-1 subacutely into the cervical spinal cord showed better performance in non-human primates with SCI [77]. NSPCs transfected with Galectin-1 using a lentivirus and transplanted into the spinal cord of non-human primates after 9 days of injury enhanced functional recovery and neurite outgrowth

[77]. In another study, hNSCs were treated with heparin, laminin, and bFGF and then transplanted into the spinal cord lesion site of the rat. Cells showed an improved survival rate, neuronal differentiation, and oligo-dendroglia differentiation compared to untreated cells [44]. Most experiments involving spinal cord therapy involve rodent stem cells rather than human stem cells. Human stem cells have certain drawbacks including ethical concerns and difficulty in growing.

Embryonically derived NSCs can be isolated from various parts of the CNS between embryonic day 13 and 16. They can be expanded as neurospheres and injected into the injured spinal cord, and they are capable of self-renewal in the presence of epidermal growth factor (EGF) or/and fibroblast growth factor (FGF) [78]. These two types of growth factors have their own receptors, which are expressed in the cytoplasm and nucleus of the NSCs and promote growth of NSPCs in both in vivo and in vitro environments [79, 80].

4.1.3 Induced Pluripotent Stem Cells

Induced pluripotent stem cells are generated from adult somatic cells that have been genetically reprogrammed to an embryonic stem cell-like state by introducing genes that are important for maintaining the essential properties of embryonic stem cells [81]. In comparison to ESCs, iPSCs have regenerative therapeutic potential without many ethical problems. The use of iPSCs avoids a number of concerns such as formation of tumor, low engraftment rates, and immune rejections, which are caused by the use of hESCs. iPSCs are induced from adult somatic cells through a process that restores pluripotency. They were first derived from the mouse somatic cell with retroviral transduction of transcription factors c-Myc, Sox2 (sex-determining region Ybox2), Klf4 (Kruppel-like factor),) and Oct ¾ (octamer-4) into the cell [82]. iPSCs can also be generated from human somatic cells [81, 83]. They can be generated from human NSCs by providing a single transcription factor (OCT4) or by delivering recombinant proteins directly [84, 85]. They can differentiate into different cell types, including NPCs, motoneurons, neurons, and glia [86, 87].

Primary skin fibroblasts can be reprogrammed to pluripotent stem cells that differentiate into neural stem cells. Being able to reprogram a patient's somatic cells into iPSCs would promote the development of a patient-specific treatment [88]. Then the differentiated cells can be transplanted into the lesion of SCI. The implanted iPSC-derived NSCs proliferate and differentiate into the neurons to establish functional connection [16]. The iPSC-derived NSCs from epithelial cells of an 86-year-old man were grafted into a rat spinal cord. Transplantation of iPSC-derived NSCs into the SCI formed a neuronal relay across the injury site, where host axons started to grow into the graft and form a synapse on the grafted cells, which in turn extend their axons to a distant target. This observation demonstrated that a neural relay formation in the lesion can function as a therapeutic treatment of SCI [16]. Host axon regeneration was observed after a relay formation across the graft on both host-to-graft and graft-to-host synapses. However, compared to rat models, neuronal maturation and myelination in the human nervous system takes a longer period of time. The absence of myelination leads to a delay in neural conduction. Thus, for the functional improvement through relay formation, human iPSC (hiPSC)-derived NSCs need to be conducted for a longer period of time [16, 89]. Also, a recent study showed that transplanted hiPSC-derived self-renewing neuroepithelial-like stem cells differentiated into neuronal progeny in the injured mouse spinal cord and restored synaptic connections, contributing to improved motor function [51].

Though NSPCs can be derived from hiPSCs in recent experiments, hiPSC differentiation to neural lineages occurs at a much lower frequency than with embryonic stem cells [90]. hESCs as well as adult stem cells have been a primary cell source for bone, cardiac, and hepatic tissue engineering. Although this has proven to be a productive area of research, the use of ESCs is limited by ethical concerns and short life spans [91]. hiPSCs have been recognized as an alternative cell source and avoid much of the ethical concerns inherent to ESCs. However, some types of iPSC-derived neural cells have an increased likelihood of tumor formation after transplantation into the CNS. Thus, safe iPSC-derived clones will need to be screened and selected [49, 92]. Experimental studies using preselected "safe" iPSC-derived neurospheres transplanted subacutely after contusion SCI showed remyelination, axonal outgrowth of serotonergic fibers, and promotion of locomotor recovery [49]. In contrast, transplantation of "unsafe" iPSC-derived neurospheres resulted in robust teratoma formation and sudden loss of locomotor function [49]. Transplantation of murine iPSC-derived astrocytes into SCI rats resulted in allodynia [52]. Recently, Okano and colleagues grafted hiPSC-derived neurospheres into an injured mouse spinal cord and demonstrated improved locomotor recovery with synapse formation between host and grafted cells, expression of neurotrophic factors, angiogenesis, axonal regrowth, and increased myelination [50]. No tumor formation occurred in the grafted mice with the preselected clones.

4.1.4 Mesenchymal Stem Cells

Mesenchymal stem cells were first isolated from bone marrow [93, 94]. They can also be isolated from other tissues such as adult adipose tissue, the umbilical cord, skin tissues, and also recently from olfactory tissue [95, 96]. MSCs are self-renewing multipotent progenitor cells that can differentiate into osteoblasts, chondrocytes, and adipocytes in vitro [97]. They are biologically safe for transplantation in patients and have been used for cell therapy [98]. Different from embryonic stem cells, MSCs can be used for autologous transplantation without immune response and ethical problems [99]. MSCs have demonstrated the ability to support axonal growth and remyelination [100]. MSCs can also protect the injured CNS tissue from apoptotic cells because they secrete anti-inflammatory, trophic factors and anti-apoptotic molecules and promote angiogenesis [101–104]. The isolation procedure of MSCs is simple and easy [105], cell expansion can be carried out within a short period of time

[106], and the cells can be preserved conveniently [107]. MSCs can be genetically modified to promote regeneration of neurons [108] and also their own survival [109].

Human umbilical cord-derived MSCs have been used to treat CNS injury and related diseases. In one study, umbilical cord stem cells were transplanted into the rat spinal cord 1 week after injury, and the transplanted cells were found to differentiate into neurons, oligodendrocytes, and astrocytes. In addition, umbilical cord stem cells facilitated the synthesis of myelin basic protein and proteolipid protein that promoted the axon myelination process [110]. When human umbilical cord blood-derived MSCs were injected into shiverer mice, the cells survived and migrated in vivo and myelinated the denuded axons of shiverer mice brain [111].

The use of MSCs obtained from Wharton's jelly of umbilical cord has no ethical concerns because they are removed after birth. It was found that the MSCs derived from human umbilical Wharton's jelly can differentiate into Schwann cell-like cells and produce neurotrophic factors such as nerve growth factor (NGF), neurotrophin-3 (NT-3), and BDNF, which stimulate neurite growth in vitro [112]. In addition, they can secrete other different types of growth factors such as insulin-like growth factor (IGF), transforming growth factor (TGF), FGF, BDGF, vascular endothelial growth factors (VEGF), and granulocyte-macrophage colony stimulating factor (GCSF) [113–116]. Furthermore, it was found that these cells can produce NT-3, bFGF, human neutrophil-activating protein-2, vascular endothelial growth factor receptor 3 (VEGFR3), and glucocorticoid-induced tumor necrosis factor receptor, which have aided in spinal cord repair. Human umbilical mesenchymal stem cells (HUMSCs) isolated from Wharton's jelly of the umbilical cord were transplanted into the rat spinal cord with complete transection. Significant locomotion improvements were observed in all rats receiving HUMSCs 3 weeks after transplantation [53]. The effects of HUMSCs were studied by transplanting them into completely transected rat and found they survived for 16 weeks.

Bone marrow-derived MSCs (BMSCs) differentiate not only into mesodermal cell lineage but also into neuronal and glial cell lineage under certain experimental conditions. Studies have shown that the presence of a chronic scar in spinal cord injury can block axonal growth. This chronic scar is due to the presence of extracellular matrix-associated inhibitors and the chronic increase in number of astrocytes in the injury site. BMSCs were genetically modified to express NT-3 and then grafted into the adult rat without removing the chronic scar. Results showed the penetration of axons through the chronic scar in the lesion with the expression NT-3. In the presence of growth stimulating factors, axonal regeneration was observed in chronically injured spinal cords [6]. Furthermore, BMSCs may provide growth factors and enhance axonal elongation along the lesion site [117]. Studies have shown that the transplantation of BMSCs resulted in decreased demyelination, suppression of neuroinhibitory molecules, and enhanced axonal regeneration [107]. In addition, the transplantation of BMSCs into SCI rat models led to partial improvement in neural regeneration with significant motor function restoration [54, 118]. In similar studies, the MSCs isolated from rat [56, 119], canine [120], and human [71] were differentiated as neural cells and then transplanted into the wounded spinal cord. The transplantation resulted in an increase in neural regeneration and motor function recovery, and a reduction in inflammatory cells. The transplantation of rat [55] and human [121]-derived MSCs along with Schwann cells into an injured spinal cord of a rat model resulted in limited scar formation and enhanced axonal remyelination. Studies have also proposed that the MSCs can act as a "guiding cue" for the regrowth of axons in the injured spinal cord [122]. The transplantation of MSCs provided both guidance and a cellular bridge across the lesion site that was populated with immature astrocytes and nerve fibers [122]. It was reported that human MSCs express cell adhesion molecule and receptors such as ninjurin 1 and 2, netrin 4, Robo1, and Robo2, which function in MSC and neuronal interaction through axon guidance and hence enhance axonal regeneration [123, 124].

Adipose-derived mesenchymal stem cells (ADSCs) can be isolated by liposuction or abdominoplasty [125] and have demonstrated multilineage differentiation potential [126]. It was reported that the ADSCs were differentiated into neural-like cells in vitro [127]. It was also reported that the ADSCs cells can enhance axonal regeneration after peripheral injury, protect neurons, and promote angiogenesis by providing neurotrophic factors such as BDNF, NT-3, and NT-4 [128–130, 131]. Several studies have been done in vivo using ADSC-derived SC-like cells for nerve repair. The graft of fibrin neural conduits seeded with these cells in injured sciatic nerve has significantly improved nerve regeneration compared with empty neural conduits [128]. In another study, SC-like cells differentiated from bone marrow-derived mesenchymal stem cells and SC-like cells differentiated from adipose-derived stem cells (ADSCs) were seeded in commercially available collagen conduits (NeuraGen(®) nerve guides). These prepared neural conduits were used to bridge the severed rat sciatic nerve. However, the neural conduit-seeded stem cell-derived SC-like cells did not significantly improve nerve regeneration [132]. Another study was also done to promote nerve growth by neuronal loss reduction through the supply of growth factors [133]. Studies have shown the differentiation of ADSCs into Schwann cells in the presence of growth factors such as platelet-derived growth factor, fibroblast growth factor, and glial growth factor [134–137]. The ADSCs-Schwann cells promoted regeneration and remyelination of the injured nerves. The transplantation of ADSCs in animal models promoted the sprouting of spared axons across the injury site and plasticity after spinal cord injury.

4.2 Biomaterial and Stem Co-Transplantation in Neural Regeneration

Biomaterials have offered an effective carrier means for stem cell transplantation following SCI. Various complications and pathophysiological concerns that arise during axonal regeneration following SCI may be overcome by the co-transplantation of biomaterial scaffolds fabricated from natural and synthetic polymers. Biomaterial scaffolds have proven effective at ameliorating cell death following transplantation and preventing scar formation.

Biomaterial scaffolds delivered to the injury site in the presence of growth factors such as NT-3 and PDGF [138–140] have been shown to promote spinal cord regeneration. A related study transplanted a collagen scaffold composed of a three-dimensional (3D) biodegradable matrix with hESC-derived neural progenitors into an SCI rat model [30]. The transplantation of hESC-derived progenitor cells has been delivered by extracellular matrix components such as collagen [30], fibronectin, and laminin [141] for SCI treatment. Collagen filaments function as a bridge, enabling neural regeneration when grafted to the spinal cord of injured rats. In addition, collagen conduits were used as a reservoir for NT-3 gene delivery, which improved the rate of axon regeneration 1 month following spinal cord injury in the rat model [142]. The use of fibrin scaffolds to deliver NT-3 to SCI rats demonstrated robust neuronal fiber growth while significant improvement in motor recovery was not observed [143]. Transplantation of ESCs does bear the risk of teratoma formation. To solve the problem of tumor formation, prior to ESCs-based clinical trials, many studies have been performed. One such study demonstrated that if ESCs are pre-differentiated prior to grafting, teratoma formation is inhibited. The subacute transplantation of pre-differentiated mouse ESCs into the injured rat spinal cord enabled cells to differentiate into neurons and glia accompanied by partial functional recovery [38]. In addition, the co-transplantation of ESCs with MSCs resulted in axonal regeneration, myelination, and angiogenesis [37]. ESCs transplanted with BMSCs have been used to prevent the formation of teratomas as well. BMSCs produce different types of neurotrophic factors including nerve growth factor, glial cell line-derived neurotrophic factor (GDNF), and BDNF, all of which induce undifferentiated ES cells to differentiate into a neuronal lineage that helps prevent tumor formation [37].

Gels, sponges, and conduits are different types of artificial scaffolds that are also used for SCI experiments. The application of synthetic neural conduits across the lesion site promotes axonal regeneration and also reduces the glial cell barrier. Implantation of extramedullary chitosan channels seeded with NSPCs derived from adult rat spinal cord after complete spinal cord transection created a large tissue bridge. In one study, synthetic guidance channels, chitosan channels, were applied to promote axonal regeneration across the lesion site. Radial glial cells from neural stem cells may migrate into the grafted chitosan channels and regenerate axons in the lesion site of spinal cord injury [144]. Hyaluronic acid has been used to develop an implantable scaffold [47, 145–148], and other components of the brain's extracellular matrix, such as laminin, can enhance the effectiveness of the scaffold [141, 149]. One recent study used a composite scaffold of salmon fibrin, hyaluronic acid, and laminin to provide sustained support to grafted NSCs [148]. The fibrin scaffolds generated from salmon fibrinogen promoted greater human neural stem cell proliferation than those of mammals. The combination scaffold, which included salmon fibrin with interspersed networks of hyaluronic acid and laminin, polymerized more effectively than fibrin alone, generating a compliant hydrogel with physical properties tuned to brain tissue.

Although research has proven promising, stem cell grafting is impeded by the low survival rate of grafted cells at the site of spinal cord lesion. One primary goal of current studies is to optimize the survival and to direct the differentiation of grafted neural stem cells. To address the low survival rate (2–4%) of transplanted cells, transplantation vehicles designed with enhanced response to injury-induced signals such as the chemokine stromal cell-derived factor-1a (SDF-1a), a chemotactic signal present following traumatic brain injury, are being developed [145]. Hyaluronic acid hydrogels containing embedded BDNF for controlled delivery have been shown to effectively create microenvironments for neuronal survival and differentiation [146, 147, 150]. A sodium hyaluronate scaffold embedded with a bFGF-controlled releasing system (bFGF-CRS) was shown to induce neural stem cell differentiation into mature neurons at a high percentage (82 ± 1.528%) within 2 weeks [47]. The long-term activation of bFGF receptors up-regulated ERK/MAPK signal pathways, which promoted the development of presynaptic and postsynaptic structures among induced neuronal cells. Biomaterial vehicles developed from highly tunable ionic diblock copolypeptide hydrogel (DCH) have demonstrated the sustained release of both hydrophilic and hydrophobic molecules within CNS [151]. Non-ionic and thermoresponsive forms of DCH have demonstrated remarkable cytocompatibility, leading to a three-fold increase in the survival of grafted NSCs into healthy CNS, with significant integration and distribution in non-neural lesion cores supporting the regrowth of host nerve fibers in injured CNS. The directed differentiation of ESC-derived neural progenitor cells into neurons and oligodendrocytes by using embedded growth factors (GF), heparin-binding delivery systems [152], and self-assembly/designer models [48, 153, 154] has greatly mitigated the poor control of cell differentiation following transplantation. Improving the efficacy of neural stem cell transplantation has led to significant developments in biomaterial applications, especially in scaffold production. With dramatic improvements in graft survival following implantation, tracking, and imaging methods, and the controlled differentiation and extension at the site of injury, the field of biomaterials as related to neural stem cells is a burgeoning area of research and development.

The survival, differentiation, and axonal extension of grafted cells can be tracked by development of the imaging system [46]. Nanoparticles used for cell tracking have proven to be powerful tools to illustrate cellular migration and homing functions of transplanted stem cells [155]. Maghemite nanoparticles were coated with poly (L-lysine) to improve biocompatibility as a cell tracer and scored high in cell-labeling efficiency. The labeling efficiency, viability, and proliferation of neural stem cells were determined using a methyl thiazolyl tetrazolium assay and calcein acetoxymethyl ester/propidium iodide assay. The use of a bicistronic vector with neuron gene tubb3 promoter is one of the many new tools used to noninvasively monitor the process of neural stem cell differentiation following transplantation in vivo [156]. Spatiotemporal imaging results demonstrated multiplexed migration, proliferation, and differentiation of the transplanted NSCs in mice models.

Special attention is required to avoid potential xenopathogenic transmission and mitigating immune rejection through animal-derived products that may need to be exposed to iPSCs. The development of an effective feeder cell-free culture environment may be necessary to reduce batch variation and improve scalability [157]. The optimization of an extracellular matrix (ECM) system for iPSC culture is a priority of investigation and nanofabrication by electrospinning has generated a promising substrate for iPSC culturing. The development of biocompatible polyacrylonitrile (PAN) fibers have been used to develop a non-cell-based feeder layer as a means to address the concerns of xenogenic contamination and low productivity of existing feeder cell preparation [158].

iPSCs have been investigated in bone tissue engineering. Patient-derived human gingival fibroblasts (hGFs) were used to investigate the osteogenic differentiation after the cells were differentiated into hiPSCs and seeded on nanohydroxyapatite (nHA) scaffolds. nHA is an important part of natural bones, and when combined with chitosan and gelatin (HCG scaffold), the hybrid acts as an important regulator and may promote the osteogenic marker expression of hiPSCs. This study showed that the sphere-nHA/CG significantly increased hiPSCs proliferation and osteogenic differentiation in vitro [159]. The co-transplantation of hiPSCs and sphere-shaped nHA/CG scaffolds led to the generation of large bones in vivo.

The essential application of iPSCs in cardiovascular applications has been investigated in the study of polymeric bio-scaffolds for myocardial patch repair of post-myocardial infarction damage [160]. The effect of novel polymer, pol3-hydroxybutyrate-co-3-hydroxyhexanoate on mouse-iPSC proliferation and differentiation has been tested in in vitro studies. Cell proliferation and cell differentiation studies illustrate the advantages of three-dimensional cultures versus two-dimensional cultures. Bioactive cardiac implants require special attention as to the choice of biomaterial and stem cell source to avoid disruption to cardiac syncytium and the de-synchronization of cardiac rhythm following intra-cardiac implantation of contractile cells [91, 161]. RAD16-I is a biomaterial that has been proven as a commonly used material for cardiac implants. Human iPSCs cultured within the RAD16-I bio-scaffold showed marked up-regulation of genes associated with early cardiogenesis. However, the molecules related with mature cardiomyocytes such as connexin 43 or troponin I proteins were not detected, and the cardiac contraction in the constructs was observed [161]. In one interesting study, common, commercially available paper materials (print paper, chromatography paper, and nitrocellulose membrane) were used to fabricate a paper-based array for the culture, proliferation, and cardiac differentiation of human iPSCs. Grown on these paper substrates, the hiPSCs presented with three-dimensional morphology and pluripotent properties. The hiPSCs were differentiated into functional cardiac tissues on paper, and the cardiac tissue retained its functionality with a beat frequency of 40–70 beats per minute lasting for 3 months [162].

The development of hepatic cells derived from hiPSCs has been a useful tool in the design of viable drug-screening processes [163–165]. The basic liver structure is the hexagonal lobule units formed by hepatocytes along with those supporting cells of mesodermal and endodermal origins. The expeditious 3D digital bioprinting can generate engineered biomimetic hepatic tissue [163, 166]. The 3D printing technology has been used to fabricate a hydrogel-based tri-culture model by embedding hiPSC-derived hepatic progenitor cells (hiPSC-HPCs), human umbilical vein endothelial cells, and adipose-derived stem cells in the hydrogel with a microscale hexagonal structure. In comparison to 2D monolayer models and 3D HPC-only models, the engineered tri-culture model showed functional improvements with the enhanced morphological structure, increased liver-specific gene expression, and increased induction of cytochrome P450. The study demonstrated that bioprinting technology is a viable and gentle enough procedure to print hiPSCs in a 3D structure while maintaining both pluripotency and differentiating capacity [166].

Although transplantation of iPSCs alone has demonstrated efficacy in the repair of wounded spinal cord, limited research has been undertaken to study the effect of co-transplantation of iPSCs and biomaterial scaffolds on spinal cord regeneration. One in vitro study showed that iPSCs were efficiently differentiated into neurons when they were grown in a chitin-alginate microfibrous scaffold [167]. It was reported that the differentiation efficiency of iPSCs into neurons in fibrin matrices was low compared with embryonic stem cells. A recent study optimized the condition of iPSC differentiation in fibrin. After the iPSC-derived embryoid bodies (EBs) in fibrin were generated using a six-day 2−/4+ protocol with soluble retinoic acid and the small-molecule sonic hedgehog agonist purmorphamine in the last four days, the EBs yielded a higher percentage of neurons compared to those from the 4−/4+ protocol for both iPSCs and ESCs. This study demonstrated the potential application of fibrin-based cell delivery for the treatment of wounded spinal cord [88]. In one in vivo study, after human iPSCs derived from skin fibroblasts were differentiated into neural crest stem cells (NCSCs) in vitro, they were mixed with hydrogel and seeded on nanofibrous scaffolds for transplantation into the fetal lamb spinal cord with surgically created myelomeningocele (MMC). Thirty days after transplantation, the grafted cells were labeled with a neurofilament antibody that indicated the neuron differentiation of the cells in the host tissue [168]. In another study, injectable hydrogel comprising hyaluronan and methylcellulose seeded with iPSC-derived oligodendrocyte precursor cells was implanted into the rat spinal cord with a moderate clip-compression injury. After transplantation in the hydrogel, most cells differentiated to a glial phenotype. This study indicated that hydrogel promoted cell survival and attenuated teratoma formation due to the fact that the hydrogel promoted cell differentiation [169].

From the industrial to the clinical, human-induced pluripotent stem cells have found many biomaterial applications within the field of bioengineering. Novel protocols are being developed to generate bone, cardiac, and hepatic tissue, with scaffold models ranging from a sheet of paper to three-dimensional self-assembling constructs to push this field into a new and exciting frontier.

References

1. Tator C. Epidemiology and general characteristics of the spinal cord injury patient. In: Benzel E, Tator CH, editors. Contemporary Management of Spinal Cord Injury. Park Ridge: American Association of Neurological Surgeons; 1995. p. 9–13.
2. Gaudet AD, Popovich PG, Ramer MS. Wallerian degeneration: gaining perspective on inflammatory events after peripheral nerve injury. J Neuroinflammation. 2011;8(1):110.
3. Trapp BD, Hauer P, Lemke G. Axonal regulation of myelin protein mRNA levels in actively myelinating Schwann cells. J Neurosci. 1988;8(9):3515–21.

4. Stoll G, Griffin J, Li CY, Trapp B. Wallerian degeneration in the peripheral nervous system: participation of both Schwann cells and macrophages in myelin degradation. J Neurocytol. 1989;18(5):671–83.
5. Dawson MR, Levine JM, Reynolds R. Mini-review-NG2-expressing cells in the central nervous system: are they oligodendroglial progenitors? J Neurosci Res. 2000;61(5):471–9.
6. Lu P, Jones LL, Tuszynski MH. Axon regeneration through scars and into sites of chronic spinal cord injury. Exp Neurol. 2007;203(1):8–21.
7. Silver J, Miller JH. Regeneration beyond the glial scar. Nat Rev Neurosci. 2004;5(2):146–56.
8. Jones LL, Tuszynski MH. Spinal cord injury elicits expression of keratan sulfate proteoglycans by macrophages, reactive microglia, and oligodendrocyte progenitors. J Neurosci. 2002;22(11):4611–24.
9. Tang X, Davies JE, Davies SJ. Changes in distribution, cell associations, and protein expression levels of NG2, neurocan, phosphacan, brevican, versican V2, and tenascin-C during acute to chronic maturation of spinal cord scar tissue. J Neurosci Res. 2003;71(3):427–44.
10. Viader A, Chang L-W, Fahrner T, Nagarajan R, Milbrandt J. MicroRNAs modulate Schwann cell response to nerve injury by reinforcing transcriptional silencing of dedifferentiation-related genes. J Neurosci. 2011;31(48):17358–69.
11. Stenudd M, Sabelström H, Frisén J. Role of endogenous neural stem cells in spinal cord injury and repair. JAMA Neurol. 2015;72(2):235–7.
12. De Feo D, Merlini A, Laterza C, Martino G. Neural stem cell transplantation in central nervous system disorders: from cell replacement to neuroprotection. Curr Opin Neurol. 2012;25(3):322–33.
13. Dalous J, Larghero J, Baud O. Transplantation of umbilical cord-derived mesenchymal stem cells as a novel strategy to protect the central nervous system: technical aspects, preclinical studies, and clinical perspectives. Pediatr Res. 2012;71(4–2):482–90.
14. Cusimano M, Biziato D, Brambilla E, Donegà M, Alfaro-Cervello C, Snider S, et al. Transplanted neural stem/precursor cells instruct phagocytes and reduce secondary tissue damage in the injured spinal cord. Brain. 2012;135(2):447–60.
15. Yang N, Zuchero JB, Ahlenius H, Marro S, Ng YH, Vierbuchen T, et al. Generation of oligodendroglial cells by direct lineage conversion. Nat Biotechnol. 2013;31(5):434–9.
16. Lu P, Woodruff G, Wang Y, Graham L, Hunt M, Wu D, et al. Long-distance axonal growth from human induced pluripotent stem cells after spinal cord injury. Neuron. 2014;83(4):789–96.
17. Evans MJ, Kaufman MH. Establishment in culture of pluripotential cells from mouse embryos. Nature. 1981;292(5819):154–6.
18. Martin GR. Isolation of a pluripotent cell line from early mouse embryos cultured in medium conditioned by teratocarcinoma stem cells. Proc Natl Acad Sci. 1981;78(12):7634–8.
19. McDonald JW, Becker D, Holekamp TF, Howard M, Liu S, Lu A, et al. Repair of the injured spinal cord and the potential of embryonic stem cell transplantation. J Neurotrauma. 2004;21(4):383–93.
20. Keirstead HS, Nistor G, Bernal G, Totoiu M, Cloutier F, Sharp K, et al. Human embryonic stem cell-derived oligodendrocyte progenitor cell transplants remyelinate and restore locomotion after spinal cord injury. J Neurosci. 2005;25(19):4694–705.
21. Sharp J, Frame J, Siegenthaler M, Nistor G, Keirstead HS. Human embryonic stem cell-derived oligodendrocyte progenitor cell transplants improve recovery after cervical spinal cord injury. Stem Cells. 2010;28(1):152–63.
22. Waxman S, Utzschneider D, Kocsis J. Enhancement of action potential conduction following demyelination: experimental approaches to restoration of function in multiple sclerosis and spinal cord injury. Prog Brain Res. 1994;100:233–43.
23. Jeffery N, Blakemore W. Locomotor deficits induced by experimental spinal cord demyelination are abolished by spontaneous remyelination. Brain. 1997;120(1):27–37.
24. Jeffery N, Crang A, O'leary M, Hodge S, Blakemore W. Behavioural consequences of oligodendrocyte progenitor cell transplantation into experimental demyelinating lesions in the rat spinal cord. Eur J Neurosci. 1999;11(5):1508–14.

References

25. Pintér S, Gloviczki B, Szabó A, Márton G, Nógrádi A. Increased survival and reinnervation of cervical motoneurons by riluzole after avulsion of the C7 ventral root. J Neurotrauma. 2010;27(12):2273–82.
26. Chen J, Bernreuther C, Dihné M, Schachner M. Cell adhesion molecule l1-transfected embryonic stem cells with enhanced survival support regrowth of corticospinal tract axons in mice after spinal cord injury. J Neurotrauma. 2005;22(8):896–906.
27. Cui Y-F, Xu J-C, Hargus G, Jakovcevski I, Schachner M, Bernreuther C. Embryonic stem cell-derived L1 overexpressing neural aggregates enhance recovery after spinal cord injury in mice. PLoS One. 2011;6(3):e17126.
28. Galichet C, Guillemot F, Parras CM. Neurogenin 2 has an essential role in development of the dentate gyrus. Development. 2008;135(11):2031–41.
29. Perrin FE, Boniface G, Serguera C, Lonjon N, Serre A, Prieto M, et al. Grafted human embryonic progenitors expressing neurogenin-2 stimulate axonal sprouting and improve motor recovery after severe spinal cord injury. PLoS One. 2010;5(12):e15914.
30. Hatami M, Mehrjardi NZ, Kiani S, Hemmesi K, Azizi H, Shahverdi A, et al. Human embryonic stem cell-derived neural precursor transplants in collagen scaffolds promote recovery in injured rat spinal cord. Cytotherapy. 2009;11(5):618–30.
31. Rossi SL, Nistor G, Wyatt T, Yin HZ, Poole AJ, Weiss JH, et al. Histological and functional benefit following transplantation of motor neuron progenitors to the injured rat spinal cord. PLoS One. 2010;5(7):e11852.
32. Kerr CL, Letzen BS, Hill CM, Agrawal G, Thakor NV, Sterneckert JL, et al. Efficient differentiation of human embryonic stem cells into oligodendrocyte progenitors for application in a rat contusion model of spinal cord injury. Int J Neurosci. 2010;120(4):305–13.
33. Erceg S, Ronaghi M, Oria M, García Roselló M, Aragó MAP, Lopez MG, et al. Transplanted oligodendrocytes and motoneuron progenitors generated from human embryonic stem cells promote locomotor recovery after spinal cord transection. Stem Cells. 2010;28(9):1541–9.
34. Salehi M, Pasbakhsh P, Soleimani M, Abbasi M, Hasanzadeh G, Modaresi MH, et al. Repair of spinal cord injury by co-transplantation of embryonic stem cell-derived motor neuron and olfactory ensheathing cell. Iran Biomed J. 2009;13(3):125.
35. All AH, Bazley FA, Gupta S, Pashai N, Hu C, Pourmorteza A, et al. Human embryonic stem cell-derived oligodendrocyte progenitors aid in functional recovery of sensory pathways following contusive spinal cord injury. PLoS One. 2012;7(10):e47645.
36. Faulkner J, Keirstead HS. Human embryonic stem cell-derived oligodendrocyte progenitors for the treatment of spinal cord injury. Transpl Immunol. 2005;15(2):131–42.
37. Matsuda R, Yoshikawa M, Kimura H, Ouji Y, Nakase H, Nishimura F, et al. Cotransplantation of mouse embryonic stem cells and bone marrow stromal cells following spinal cord injury suppresses tumor development. Cell Transplant. 2009;18(1):39–54.
38. McDonald JW, Liu X-Z, Qu Y, Liu S, Mickey SK, Turetsky D, et al. Transplanted embryonic stem cells survive, differentiate and promote recovery in injured rat spinal cord. Nat Med. 1999;5(12):1410–2.
39. Glazova M, Pak ES, Moretto J, Hollis S, Brewer KL, Murashov AK. Pre-differentiated embryonic stem cells promote neuronal regeneration by cross-coupling of BDNF and IL-6 signaling pathways in the host tissue. J Neurotrauma. 2009;26(7):1029–42.
40. Hofstetter CP, Holmström NA, Lilja JA, Schweinhardt P, Hao J, Spenger C, et al. Allodynia limits the usefulness of intraspinal neural stem cell grafts; directed differentiation improves outcome. Nat Neurosci. 2005;8(3):346–53.
41. Wang G, Ao Q, Gong K, Zuo H, Gong Y, Zhang X. Synergistic effect of neural stem cells and olfactory ensheathing cells on repair of adult rat spinal cord injury. Cell Transplant. 2010;19(10):1325–37.
42. Mothe AJ, Tator CH. Transplanted neural stem/progenitor cells generate myelinating oligodendrocytes and Schwann cells in spinal cord demyelination and dysmyelination. Exp Neurol. 2008;213(1):176–90.

43. Hwang DH, Kim BG, Kim EJ, Lee SI, Joo IS, Suh-Kim H, et al. Transplantation of human neural stem cells transduced with Olig2 transcription factor improves locomotor recovery and enhances myelination in the white matter of rat spinal cord following contusive injury. BMC Neurosci. 2009;10(1):117.
44. Tarasenko YI, Gao J, Nie L, Johnson KM, Grady JJ, Hulsebosch CE, et al. Human fetal neural stem cells grafted into contusion-injured rat spinal cords improve behavior. J Neurosci Res. 2007;85(1):47–57.
45. Yan J, Xu L, Welsh AM, Hatfield G, Hazel T, Johe K, et al. Extensive neuronal differentiation of human neural stem cell grafts in adult rat spinal cord. PLoS Med. 2007;4(2):e39.
46. Lu P, Graham L, Wang Y, Wu D, Tuszynski M. Promotion of survival and differentiation of neural stem cells with fibrin and growth factor cocktails after severe spinal cord injury. J Vis Exp. 2014;(89):e50641-e.
47. Duan H, Li X, Wang C, Hao P, Song W, Li M, et al. Functional hyaluronate collagen scaffolds induce NSCs differentiation into functional neurons in repairing the traumatic brain injury. Acta Biomater. 2016;45:182–95.
48. Cheng TY, Chen MH, Chang WH, Huang MY, Wang TW. Neural stem cells encapsulated in a functionalized self-assembling peptide hydrogel for brain tissue engineering. Biomaterials. 2013;34(8):2005–16.
49. Tsuji O, Miura K, Okada Y, Fujiyoshi K, Mukaino M, Nagoshi N, et al. Therapeutic potential of appropriately evaluated safe-induced pluripotent stem cells for spinal cord injury. Proc Natl Acad Sci. 2010;107(28):12704–9.
50. Nori S, Okada Y, Yasuda A, Tsuji O, Takahashi Y, Kobayashi Y, et al. Grafted human-induced pluripotent stem-cell–derived neurospheres promote motor functional recovery after spinal cord injury in mice. Proc Natl Acad Sci. 2011;108(40):16825–30.
51. Fujimoto Y, Abematsu M, Falk A, Tsujimura K, Sanosaka T, Juliandi B, et al. Treatment of a mouse model of spinal cord injury by transplantation of human induced pluripotent stem cell-derived long-term self-renewing neuroepithelial-like stem cells. Stem Cells. 2012;30(6):1163–73.
52. Hayashi K, Hashimoto M, Koda M, Naito AT, Murata A, Okawa A, et al. Increase of sensitivity to mechanical stimulus after transplantation of murine induced pluripotent stem cell–derived astrocytes in a rat spinal cord injury model: Laboratory investigation. J Neurosurg Spine. 2011;15(6):582–93.
53. Yang C-C, Shih Y-H, Ko M-H, Hsu S-Y, Cheng H, Fu Y-S. Transplantation of human umbilical mesenchymal stem cells from Wharton's jelly after complete transection of the rat spinal cord. PLoS One. 2008;3(10):e3336.
54. Osaka M, Honmou O, Murakami T, Nonaka T, Houkin K, Hamada H, et al. Intravenous administration of mesenchymal stem cells derived from bone marrow after contusive spinal cord injury improves functional outcome. Brain Res. 2010;1343:226–35.
55. Ban D-X, Ning G-Z, Feng S-Q, Wang Y, Zhou X-H, Liu Y, et al. Combination of activated Schwann cells with bone mesenchymal stem cells: the best cell strategy for repair after spinal cord injury in rats. Regen Med. 2011;6(6):707–20.
56. Cho S-R, Kim YR, Kang H-S, Yim SH, Park C-i, Min YH, et al. Functional recovery after the transplantation of neurally differentiated mesenchymal stem cells derived from bone barrow in a rat model of spinal cord injury. Cell Transplant. 2009;18(12):1359–68.
57. Kumagai G, Okada Y, Yamane J, Nagoshi N, Kitamura K, Mukaino M, et al. Roles of ES cell-derived gliogenic neural stem/progenitor cells in functional recovery after spinal cord injury. PLoS One. 2009;4(11):e7706.
58. Penkowa M, Camats J, Hadberg H, Quintana A, Rojas S, Giralt M, et al. Astrocyte-targeted expression of interleukin-6 protects the central nervous system during neuroglial degeneration induced by 6-aminonicotinamide. J Neurosci Res. 2003;73(4):481–96.
59. Zhang YW, Denham J, Thies RS. Oligodendrocyte progenitor cells derived from human embryonic stem cells express neurotrophic factors. Stem Cells Dev. 2006;15(6):943–52.
60. Reynolds BA, Weiss S. Generation of neurons and astrocytes from isolated cells of the adult mammalian central nervous system. Science. 1992;255(5052):1707–10.

References

61. Temple S. Division and differentiation of isolated CNS blast cells in microculture. 1989.
62. Gage FH. Mammalian neural stem cells. Science. 2000;287(5457):1433–8.
63. Reubinoff BE, Itsykson P, Turetsky T, Pera MF, Reinhartz E, Itzik A, et al. Neural progenitors from human embryonic stem cells. Nat Biotechnol. 2001;19(12):1134–40.
64. Coutts M, Keirstead HS. Stem cells for the treatment of spinal cord injury. Exp Neurol. 2008;209(2):368–77.
65. Fitch MT, Silver J. CNS injury, glial scars, and inflammation: Inhibitory extracellular matrices and regeneration failure. Exp Neurol. 2008;209(2):294–301.
66. Johansson CB, Momma S, Clarke DL, Risling M, Lendahl U, Frisén J. Identification of a neural stem cell in the adult mammalian central nervous system. Cell. 1999;96(1):25–34.
67. Meletis K, Barnabé-Heider F, Carlén M, Evergren E, Tomilin N, Shupliakov O, et al. Spinal cord injury reveals multilineage differentiation of ependymal cells. PLoS Biol. 2008;6(7):e182.
68. Moreno-Manzano V, Rodríguez-Jiménez FJ, García-Roselló M, Laínez S, Erceg S, Calvo MT, et al. Activated spinal cord ependymal stem cells rescue neurological function. Stem Cells. 2009;27(3):733–43.
69. Mothe AJ, Kulbatski I, Parr A, Mohareb M, Tator CH. Adult spinal cord stem/progenitor cells transplanted as neurospheres preferentially differentiate into oligodendrocytes in the adult rat spinal cord. Cell Transplant. 2008;17(7):735–51.
70. Kulbatski I, Mothe AJ, Keating A, Hakamata Y, Kobayashi E, Tator CH. Oligodendrocytes and radial glia derived from adult rat spinal cord progenitors: morphological and immunocytochemical characterization. J Histochem Cytochem. 2007;55(3):209–22.
71. Alexanian AR, Svendsen CN, Crowe MJ, Kurpad SN. Transplantation of human glial-restricted neural precursors into injured spinal cord promotes functional and sensory recovery without causing allodynia. Cytotherapy. 2011;13(1):61–8.
72. Barnabé-Heider F, Frisén J. Stem cells for spinal cord repair. Cell Stem Cell. 2008;3(1):16–24.
73. Hsu Y-C, Lee D-C, Chiu I-M. Neural stem cells, neural progenitors, and neurotrophic factors. Cell Transplant. 2007;16(2):133–50.
74. Davies JE, Huang C, Proschel C, Noble M, Mayer-Proschel M, Davies SJ. Astrocytes derived from glial-restricted precursors promote spinal cord repair. J Biol. 2006;5(3):1.
75. Kim BG, Hwang DH, Lee SI, Kim EJ, Kim SU. Stem cell-based cell therapy for spinal cord injury. Cell Transplant. 2007;16(4):355–64.
76. Karimi-Abdolrezaee S, Eftekharpour E, Wang J, Morshead CM, Fehlings MG. Delayed transplantation of adult neural precursor cells promotes remyelination and functional neurological recovery after spinal cord injury. J Neurosci. 2006;26(13):3377–89.
77. Yamane J, Nakamura M, Iwanami A, Sakaguchi M, Katoh H, Yamada M, et al. Transplantation of galectin-1-expressing human neural stem cells into the injured spinal cord of adult common marmosets. J Neurosci Res. 2010;88(7):1394–405.
78. Tetzlaff W, Okon EB, Karimi-Abdolrezaee S, Hill CE, Sparling JS, Plemel JR, et al. A systematic review of cellular transplantation therapies for spinal cord injury. J Neurotrauma. 2011;28(8):1611–82.
79. Lee D-C, Hsu Y-C, Chung Y-F, Hsiao C-Y, Chen S-L, Chen M-S, et al. Isolation of neural stem/progenitor cells by using EGF/FGF1 and FGF1B promoter-driven green fluorescence from embryonic and adult mouse brains. Mol Cell Neurosci. 2009;41(3):348–63.
80. Türeyen K, Vemuganti R, Bowen KK, Sailor KA, Dempsey RJ. EGF and FGF-2 infusion increases post-ischemic neural progenitor cell proliferation in the adult rat brain. Neurosurgery. 2005;57(6):1254–63.
81. Takahashi K, Tanabe K, Ohnuki M, Narita M, Ichisaka T, Tomoda K, et al. Induction of pluripotent stem cells from adult human fibroblasts by defined factors. Cell. 2007;131(5):861–72.
82. Takahashi K, Yamanaka S. Induction of pluripotent stem cells from mouse embryonic and adult fibroblast cultures by defined factors. Cell. 2006;126(4):663–76.
83. Park I-H, Zhao R, West JA, Yabuuchi A, Huo H, Ince TA, et al. Reprogramming of human somatic cells to pluripotency with defined factors. Nature. 2008;451(7175):141–6.

84. Kim JB, Greber B, Araúzo-Bravo MJ, Meyer J, Park KI, Zaehres H, et al. Direct reprogramming of human neural stem cells by OCT4. Nature. 2009;461(7264):649–53.
85. Kim D, Kim C-H, Moon J-I, Chung Y-G, Chang M-Y, Han B-S, et al. Generation of human induced pluripotent stem cells by direct delivery of reprogramming proteins. Cell Stem Cell. 2009;4(6):472.
86. Dimos JT, Rodolfa KT, Niakan KK, Weisenthal LM, Mitsumoto H, Chung W, et al. Induced pluripotent stem cells generated from patients with ALS can be differentiated into motor neurons. Science. 2008;321(5893):1218–21.
87. Wernig M, Zhao J-P, Pruszak J, Hedlund E, Fu D, Soldner F, et al. Neurons derived from reprogrammed fibroblasts functionally integrate into the fetal brain and improve symptoms of rats with Parkinson's disease. Proc Natl Acad Sci. 2008;105(15):5856–61.
88. Montgomery A, Wong A, Gabers N, Willerth SM. Engineering personalized neural tissue by combining induced pluripotent stem cells with fibrin scaffolds. Biomater Sci. 2015;3(2):401–13.
89. Lu P, Wang Y, Graham L, McHale K, Gao M, Wu D, et al. Long-distance growth and connectivity of neural stem cells after severe spinal cord injury. Cell. 2012;150(6):1264–73.
90. Hu B-Y, Weick JP, Yu J, Ma L-X, Zhang X-Q, Thomson JA, et al. Neural differentiation of human induced pluripotent stem cells follows developmental principles but with variable potency. Proc Natl Acad Sci. 2010;107(9):4335–40.
91. Ji J, Tong X, Huang X, Zhang J, Qin H, Hu Q. Patient-derived human induced pluripotent stem cells from gingival fibroblasts composited with defined nanohydroxyapatite/chitosan/gelatin porous scaffolds as potential bone graft substitutes. Stem Cells Transl Med. 2016;5(1):95–105.
92. Miura K, Okada Y, Aoi T, Okada A, Takahashi K, Okita K, et al. Variation in the safety of induced pluripotent stem cell lines. Nat Biotechnol. 2009;27(8):743–5.
93. Prockop DJ. Marrow stromal cells as stem cells for nonhematopoietic tissues. Science. 1997;276(5309):71–4.
94. Friedenstein A, Deriglasova U, Kulagina N, Panasuk A, Rudakowa S, Luria E, et al. Precursors for fibroblasts in different populations of hematopoietic cells as detected by the in vitro colony assay method. Exp Hematol. 1973;2(2):83–92.
95. Tomé M, Lindsay SL, Riddell JS, Barnett SC. Identification of nonepithelial multipotent cells in the embryonic olfactory mucosa. Stem Cells. 2009;27(9):2196–208.
96. Delorme B, Nivet E, Gaillard J, Häupl T, Ringe J, Devèze A, et al. The human nose harbors a niche of olfactory ectomesenchymal stem cells displaying neurogenic and osteogenic properties. Stem Cells Dev. 2009;19(6):853–66.
97. Friedenstein AJ, Petrakova KV, Kurolesova AI, Frolova GP. Heterotopic transplants of bone marrow. Transplantation. 1968;6(2):230–47.
98. Garbossa D, Boido M, Fontanella M, Fronda C, Ducati A, Vercelli A. Recent therapeutic strategies for spinal cord injury treatment: possible role of stem cells. Neurosurg Rev. 2012;35(3):293–311.
99. Augello A, Kurth TB, De Bari C. Mesenchymal stem cells: a perspective from in vitro cultures to in vivo migration and niches. Eur Cell Mater. 2010;20:121–33.
100. Kotobuki N, Hirose M, Takakura Y, Ohgushi H. Cultured autologous human cells for hard tissue regeneration: preparation and characterization of mesenchymal stem cells from bone marrow. Artif Organs. 2004;28(1):33–9.
101. Caplan AI, Dennis JE. Mesenchymal stem cells as trophic mediators. J Cell Biochem. 2006;98(5):1076–84.
102. da Silva Meirelles L, Fontes AM, Covas DT, Caplan AI. Mechanisms involved in the therapeutic properties of mesenchymal stem cells. Cytokine Growth Factor Rev. 2009;20(5):419–27.
103. Uccelli A, Benvenuto F, Laroni A, Giunti D. Neuroprotective features of mesenchymal stem cells. Best Pract Res Clin Haematol. 2011;24(1):59–64.
104. van Poll D, Parekkadan B, Rinkes IB, Tilles AW, Yarmush ML. Mesenchymal stem cell therapy for protection and repair of injured vital organs. Cell Mol Bioeng. 2008;1(1):42–50.

References

105. Lee MW, Yang MS, Park JS, Kim HC, Kim YJ, Choi J. Isolation of mesenchymal stem cells from cryopreserved human umbilical cord blood. Int J Hematol. 2005;81(2):126–30.
106. Sekiya I, Larson BL, Smith JR, Pochampally R, Cui JG, Prockop DJ. Expansion of human adult stem cells from bone marrow stroma: conditions that maximize the yields of early progenitors and evaluate their quality. Stem Cells. 2002;20(6):530–41.
107. Malgieri A, Kantzari E, Patrizi MP, Gambardella S. Bone marrow and umbilical cord blood human mesenchymal stem cells: state of the art. Int J Clin Exp Med. 2010;3(4):248–69.
108. Liu Y, Himes BT, Murray M, Tessler A, Fischer I. Grafts of BDNF-producing fibroblasts rescue axotomized rubrospinal neurons and prevent their atrophy. Exp Neurol. 2002; 178(2):150–64.
109. Cao Q, Xu X-M, DeVries WH, Enzmann GU, Ping P, Tsoulfas P, et al. Functional recovery in traumatic spinal cord injury after transplantation of multineurotrophin-expressing glial-restricted precursor cells. J Neurosci. 2005;25(30):6947–57.
110. Dasari VR, Spomar DG, Gondi CS, Sloffer CA, Saving KL, Gujrati M, et al. Axonal remyelination by cord blood stem cells after spinal cord injury. J Neurotrauma. 2007;24(2):391–410.
111. Veeravalli KK, Dasari VR, Fassett D, Dinh DH, Rao JS. Human umbilical cord blood-derived mesenchymal stem cells upregulate myelin basic protein in shiverer mice. Stem Cells Dev. 2011;20(5):881–91.
112. Peng J, Wang Y, Zhang L, Zhao B, Zhao Z, Chen J, et al. Human umbilical cord Wharton's jelly-derived mesenchymal stem cells differentiate into a Schwann-cell phenotype and promote neurite outgrowth in vitro. Brain Res Bull. 2011;84(3):235–43.
113. Nakagami H, Maeda K, Morishita R, Iguchi S, Nishikawa T, Takami Y, et al. Novel autologous cell therapy in ischemic limb disease through growth factor secretion by cultured adipose tissue–derived stromal cells. Arterioscler Thromb Vasc Biol. 2005;25(12):2542–7.
114. Wei X, Du Z, Zhao L, Feng D, Wei G, He Y, et al. IFATS Collection: The Conditioned Media of Adipose Stromal Cells Protect Against Hypoxia-Ischemia-Induced Brain Damage in Neonatal Rats. Stem Cells. 2009;27(2):478–88.
115. Rehman J, Traktuev D, Li J, Merfeld-Clauss S, Temm-Grove CJ, Bovenkerk JE, et al. Secretion of angiogenic and antiapoptotic factors by human adipose stromal cells. Circulation. 2004;109(10):1292–8.
116. Sadat S, Gehmert S, Song Y-H, Yen Y, Bai X, Gaiser S, et al. The cardioprotective effect of mesenchymal stem cells is mediated by IGF-I and VEGF. Biochem Biophys Res Commun. 2007;363(3):674–9.
117. Wright KT, Masri WE, Osman A, Chowdhury J, Johnson WE. Concise review: bone marrow for the treatment of spinal cord injury: mechanisms and clinical applications. Stem Cells. 2011;29(2):169–78.
118. Park WB, Kim SY, Lee SH, Kim H-W, Park J-S, Hyun JK. The effect of mesenchymal stem cell transplantation on the recovery of bladder and hindlimb function after spinal cord contusion in rats. BMC Neurosci. 2010;11(1):119.
119. Pedram M, Dehghan M, Soleimani M, Sharifi D, Marjanmehr S, Nasiri Z. Transplantation of a combination of autologous neural differentiated and undifferentiated mesenchymal stem cells into injured spinal cord of rats. Spinal Cord. 2010;48(6):457–63.
120. Park S-S, Lee YJ, Lee SH, Lee D, Choi K, Kim W-H, et al. Functional recovery after spinal cord injury in dogs treated with a combination of Matrigel and neural-induced adipose-derived mesenchymal Stem cells. Cytotherapy. 2012;14(5):584–97.
121. Yan-Wu G, Yi-Quan K, Ming L, Ying-Qian C, Xiao-Dan J, Shi-Zhong Z, et al. Human umbilical cord-derived Schwann-like cell transplantation combined with neurotrophin-3 administration in dyskinesia of rats with spinal cord injury. Neurochem Res. 2011;36(5):783–92.
122. Hofstetter C, Schwarz E, Hess D, Widenfalk J, El Manira A, Prockop DJ, et al. Marrow stromal cells form guiding strands in the injured spinal cord and promote recovery. Proc Natl Acad Sci. 2002;99(4):2199–204.
123. Crigler L, Robey RC, Asawachaicharn A, Gaupp D, Phinney DG. Human mesenchymal stem cell subpopulations express a variety of neuro-regulatory molecules and promote neuronal cell survival and neuritogenesis. Exp Neurol. 2006;198(1):54–64.

124. Phinney DG, Baddoo M, Dutreil M, Gaupp D, Lai WT, Isakova IA. Murine mesenchymal stem cells transplanted to the central nervous system of neonatal versus adult mice exhibit distinct engraftment kinetics and express receptors that guide neuronal cell migration. Stem Cells Dev. 2006;15(3):437–47.
125. Strem BM, Hicok KC, Zhu M, Wulur I, Alfonso Z, Schreiber RE, et al. Multipotential differentiation of adipose tissue-derived stem cells. Keio J Med. 2005;54(3):132–41.
126. Zuk PA. The adipose-derived stem cell: looking back and looking ahead. Mol Biol Cell. 2010;21(11):1783–7.
127. Ahmadi N, Razavi S, Kazemi M, Oryan S. Stability of neural differentiation in human adipose derived stem cells by two induction protocols. Tissue Cell. 2012;44(2):87–94.
128. di Summa PG, Kingham PJ, Raffoul W, Wiberg M, Terenghi G, Kalbermatten DF. Adipose-derived stem cells enhance peripheral nerve regeneration. J Plast Reconstr Aesthet Surg. 2010;63(9):1544–52.
129. Erba P, Terenghi G, J Kingham P. Neural differentiation and therapeutic potential of adipose tissue derived stem cells. Curr Stem Cell Res Ther. 2010;5(2):153–60.
130. Dadon-Nachum M, Melamed E, Offen D. Stem cells treatment for sciatic nerve injury. Expert Opin Biol Ther. 2011;11(12):1591–7.
131. Ballermann M, Fouad K. Spontaneous locomotor recovery in spinal cord injured rats is accompanied by anatomical plasticity of reticulospinal fibers. Eur J Neurosci. 2006;23(8):1988–96.
132. di Summa PG, Kingham PJ, Campisi CC, Raffoul W, Kalbermatten DF. Collagen (NeuraGen®) nerve conduits and stem cells for peripheral nerve gap repair. Neurosci Lett. 2014;572:26–31.
133. Reid AJ, Sun M, Wiberg M, Downes S, Terenghi G, Kingham PJ. Nerve repair with adipose-derived stem cells protects dorsal root ganglia neurons from apoptosis. Neuroscience. 2011;199:515–22.
134. Kingham PJ, Kalbermatten DF, Mahay D, Armstrong SJ, Wiberg M, Terenghi G. Adipose-derived stem cells differentiate into a Schwann cell phenotype and promote neurite outgrowth in vitro. Exp Neurol. 2007;207(2):267–74.
135. Xu Y, Liu L, Li Y, Zhou C, Xiong F, Liu Z, et al. Myelin-forming ability of Schwann cell-like cells induced from rat adipose-derived stem cells in vitro. Brain Res. 2008;1239:49–55.
136. Mantovani C, Mahay D, Kingham PJ, Terenghi G, Shawcross SG, Wiberg M. Bone marrow- and adipose-derived stem cells show expression of myelin mRNAs and proteins. Regen Med. 2010;5(3):403–10.
137. Luca A, Faroni A, Downes S, Terenghi G. Differentiated adipose-derived stem cells act synergistically with RGD-modified surfaces to improve neurite outgrowth in a co-culture model. J Tissue Eng Regen Med. 2016;10(8):647–55.
138. Johnson PJ, Tatara A, Shiu A, Sakiyama-Elbert SE. Controlled release of neurotrophin-3 and platelet derived growth factor from fibrin scaffolds containing neural progenitor cells enhances survival and differentiation into neurons in a subacute model of SCI. Cell Transplant. 2010;19(1):89.
139. Willerth SM, Faxel TE, Gottlieb DI, Sakiyama-Elbert SE. The effects of soluble growth factors on embryonic stem cell differentiation inside of fibrin scaffolds. Stem Cells. 2007;25(9):2235–44.
140. Willerth SM, Rader A, Sakiyama-Elbert SE. The effect of controlled growth factor delivery on embryonic stem cell differentiation inside fibrin scaffolds. Stem Cell Res. 2008;1(3):205–18.
141. Tate CC, Shear DA, Tate MC, Archer DR, Stein DG, LaPlaca MC. Laminin and fibronectin scaffolds enhance neural stem cell transplantation into the injured brain. J Tissue Eng Regen Med. 2009;3(3):208–17.
142. Yao L, Daly W, Newland B, Yao S, Wang W, Chen B, et al. Improved axonal regeneration of transected spinal cord mediated by multichannel collagen conduits functionalized with neurotrophin-3 gene. Gene Ther. 2013;20(12):1149–57.
143. Taylor SJ, Rosenzweig ES, McDonald JW, Sakiyama-Elbert SE. Delivery of neurotrophin-3 from fibrin enhances neuronal fiber sprouting after spinal cord injury. J Control Release. 2006;113(3):226–35.

144. Nomura H, Kim H, Mothe A, Zahir T, Kulbatski I, Morshead CM, et al. Endogenous radial glial cells support regenerating axons after spinal cord transection. Neuroreport. 2010;21(13):871–6.
145. Addington C, Heffernan J, Millar-Haskell C, Tucker E, Sirianni R, Stabenfeldt S. Enhancing neural stem cell response to SDF-1α gradients through hyaluronic acid-laminin hydrogels. Biomaterials. 2015;72:11–9.
146. Wang Y, Wei YT, Zu ZH, Ju RK, Guo MY, Wang XM, et al. Combination of hyaluronic acid hydrogel scaffold and PLGA microspheres for supporting survival of neural stem cells. Pharm Res. 2011;28(6):1406.
147. Park J, Lim E, Back S, Na H, Park Y, Sun K. Nerve regeneration following spinal cord injury using matrix metalloproteinase-sensitive, hyaluronic acid-based biomimetic hydrogel scaffold containing brain-derived neurotrophic factor. J Biomed Mater Res A. 2010;93(3):1091–9.
148. Arulmoli J, Wright HJ, Phan DT, Sheth U, Que RA, Botten GA, et al. Combination scaffolds of salmon fibrin, hyaluronic acid, and laminin for human neural stem cell and vascular tissue engineering. Acta Biomater. 2016;43:122–38.
149. Jurga M, Dainiak MB, Sarnowska A, Jablonska A, Tripathi A, Plieva FM, et al. The performance of laminin-containing cryogel scaffolds in neural tissue regeneration. Biomaterials. 2011;32(13):3423–34.
150. Cho T, Ryu JK, Taghibiglou C, Ge Y, Chan AW, Liu L, et al. Long-term potentiation promotes proliferation/survival and neuronal differentiation of neural stem/progenitor cells. PLoS One. 2013;8(10):e76860.
151. Zhang S, Burda JE, Anderson MA, Zhao Z, Ao Y, Cheng Y, et al. Thermoresponsive Copolypeptide Hydrogel Vehicles for Central Nervous System Cell Delivery. ACS Biomater Sci Eng. 2015;1(8):705–17.
152. Johnson PJ, Tatara A, Shiu A, Sakiyama-Elbert SE. Controlled release of neurotrophin-3 and platelet-derived growth factor from fibrin scaffolds containing neural progenitor cells enhances survival and differentiation into neurons in a subacute model of SCI. Cell Transplant. 2010;19(1):89–101.
153. Moshayedi P, Nih LR, Llorente IL, Berg AR, Cinkornpumin J, Lowry WE, et al. Systematic optimization of an engineered hydrogel allows for selective control of human neural stem cell survival and differentiation after transplantation in the stroke brain. Biomaterials. 2016;105:145–55.
154. Wang J, Zheng J, Zheng Q, Wu Y, Wu B, Huang S, et al. FGL-functionalized self-assembling nanofiber hydrogel as a scaffold for spinal cord-derived neural stem cells. Mater Sci Eng C. 2015;46:140–7.
155. Pongrac IM, Dobrivojević M, Ahmed LB, Babič M, Šlouf M, Horák D, et al. Improved biocompatibility and efficient labeling of neural stem cells with poly (L-lysine)-coated maghemite nanoparticles. Beilstein J Nanotechnol. 2016;7(1):926–36.
156. Wang Z, Wang Y, Wang Z, Zhao J, Gutkind JS, Srivatsan A, et al. Polymeric Nanovehicle Regulated Spatiotemporal Real-Time Imaging of the Differentiation Dynamics of Transplanted Neural Stem Cells after Traumatic Brain Injury. ACS Nano. 2015;9(7):6683–95.
157. Chen Y, Zheng Y, Qiu D, Sun Y, Kuang S, Xu Y, et al. An extracellular matrix culture system for induced pluripotent stem cells derived from human dental pulp cells. Eur Rev Med Pharmacol Sci. 2015;19(21):4035–46.
158. Liu S-P, Lin C-H, Lin S-J, Fu R-H, Huang Y-C, Chen S-Y, et al. Electrospun Polyacrylonitrile-Based Nanofibers Maintain Embryonic Stem Cell Stemness via TGF-Beta Signaling. J Biomed Nanotechnol. 2016;12(4):732–42.
159. Ji J, Tong X, Huang X, Wang T, Lin Z, Cao Y, et al. Sphere-shaped nano-hydroxyapatite/chitosan/gelatin 3D porous scaffolds increase proliferation and osteogenic differentiation of human induced pluripotent stem cells from gingival fibroblasts. Biomed Mater. 2015;10(4):045005.
160. Shijun X, Junsheng M, Jianqun Z, Ping B. In vitro three-dimensional coculturing poly3-hydroxybutyrate-co-3-hydroxyhexanoate with mouse-induced pluripotent stem cells for myocardial patch application. J Biomater Appl. 2016;30(8):1273–82.

161. Puig-Sanvicens VA, Semino CE, zur Nieden NI. Cardiac differentiation potential of human induced pluripotent stem cells in a 3D self-assembling peptide scaffold. Differentiation. 2015;90(4):101–10.
162. Wang L, Xu C, Zhu Y, Yu Y, Sun N, Zhang X, et al. Human induced pluripotent stem cell-derived beating cardiac tissues on paper. Lab Chip. 2015;15(22):4283–90.
163. Ma X, Qu X, Zhu W, Li Y-S, Yuan S, Zhang H, et al. Deterministically patterned biomimetic human iPSC-derived hepatic model via rapid 3D bioprinting. Proc Natl Acad Sci. 2016;113(8):2206–11.
164. Asgari S, Moslem M, Bagheri-Lankarani K, Pournasr B, Miryounesi M, Baharvand H. Differentiation and transplantation of human induced pluripotent stem cell-derived hepatocyte-like cells. Stem Cell Rev Rep. 2013;9(4):493–504.
165. Tomotsune D, Hirashima K, Fujii M, Yue F, Matsumoto K, Takizawa-Shirasawa S, et al. Enrichment of Pluripotent Stem Cell-Derived Hepatocyte-Like Cells by Ammonia Treatment. PLoS One. 2016;11(9):e0162693.
166. Faulkner-Jones A, Fyfe C, Cornelissen D-J, Gardner J, King J, Courtney A, et al. Bioprinting of human pluripotent stem cells and their directed differentiation into hepatocyte-like cells for the generation of mini-livers in 3D. Biofabrication. 2015;7(4):044102.
167. Lu HF, Lim SX, Leong MF, Narayanan K, Toh RP, Gao S, et al. Efficient neuronal differentiation and maturation of human pluripotent stem cells encapsulated in 3D microfibrous scaffolds. Biomaterials. 2012;33(36):9179–87.
168. Saadai P, Wang A, Nout YS, Downing TL, Lofberg K, Beattie MS, et al. Human induced pluripotent stem cell-derived neural crest stem cells integrate into the injured spinal cord in the fetal lamb model of myelomeningocele. J Pediatr Surg. 2013;48(1):158–63.
169. Fuhrmann T, Tam RY, Ballarin B, Coles B, Elliott Donaghue I, van der Kooy D, et al. Injectable hydrogel promotes early survival of induced pluripotent stem cell-derived oligodendrocytes and attenuates longterm teratoma formation in a spinal cord injury model. Biomaterials. 2016;83:23–36.

Chapter 5
Electric Field-Guided Cell Migration, Polarization, and Division: An Emerging Therapy in Neural Regeneration

Li Yao and Yongchao Li

Abstract The endogenous EF has been detected in a developing neural system. Studies revealed that EF plays an important role in the development nervous system because the size, location, and developmental timing of EF influence cellular process. The significant biological effect of EF is its influence on the directional growth of neurites and cell migration. In vitro studies have demonstrated an applied direct-current EF-guided and promoted axonal growth and hippocampal neuron migration toward the cathode pole. Cell division is involved in development, wound healing, and pathology. The physiological role of EF-induced regulation of the cell division plane of neuronal cells has been shown in in vitro experiments. Although the mechanism of EF-directed axonal growth and cell migration is not fully understood, studies have found that the polarization of cells, activation of ion channels, and intracellular signaling pathways are involved in the regulation of EF-guided migration. These studies have provided direct evidence that an EF can potentially direct and enhance in vivo neural cell migration and axonal growth. The application of EF stimulation can be a novel technology for neural regeneration.

5.1 Electric Field Directing Axonal Growth

In the neurogenesis process at the embryo stage, neurons migrate to their specific target location to establish physiological connections. Extracellular guidance cues direct the extension of neuron axons that guide neuronal cell body migration. Recent studies have revealed that electrical activity was involved in the guidance of the axon extension of thalamic neurons when the axons grow toward the cortical target [1]. The electrical activity was also detected in the axons of cortical pyramidal neurons forming layer-specific connections [2].

Microtubules (MTs) and F-actin are major cytoskeletons in axons, and their interaction in the growth cone peripheral domain steers growth cone path-finding [3, 4]. Growth cones are able to detect and integrate attractive and repellent guidance cues in the extracellular matrix (ECM) and migrate with the guidance of those mol-

© The Author(s) 2018
L. Yao, *Glial Cell Engineering in Neural Regeneration*,
https://doi.org/10.1007/978-3-030-02104-7_5

ecules. The binding of guidance cues to cell membrane receptors can activate the intracellular signaling pathway that directs the motility of the growth cone [5–7].

In vitro studies have demonstrated the guidance effect of a steady direct-current electric field (DC EF) on the growth of neurites. In a small DC EF (0.1–1 V/cm), the neurites of cultured embryonic chick dorsal root ganglia explants grew toward the cathode pole [8]. Cathodal neurite growth was also observed when cultured *Xenopus* spinal neurons were subjected to an applied EF [9–13]. A DC EF-induced degeneration retraction, or absorption of cultured frog neural tube neurites into the cell body when they faced the anode, while cathode-facing neurites were spared from retraction [14]. The EF increased the growth rate of *Xenopus laevis* spinal neurites when turning toward the cathode and away from the anode [11]. In addition to a steady EF, studies have also demonstrated that pulsed EFs can direct axonal growth. *Xenopus* growth cones turned toward the cathode in response to a pulsed DC EF [15]. In steady or pulsed DC EFs, spiral ganglion neurites also grew toward the cathode when cultured on a laminin-coated surface, while the neurites on poly-D-lysine-coated slides turned toward the anode [12]. Although the alternating current (AC) field did not induce the asymmetric growth of neurites as did the DC EF, studies have shown that it can modulate the neurite extension. AC stimulation increased the PC12 cell neurite growth, and the cells showed better cell viability subjected to AC stimulation compared with that stimulated by constant current [16].

Small guanosine triphosphate (GTP)ases that interact with the cytoskeleton play a central role in regulating axonal extension and cell migration. To study the mechanism of EF-guided axonal growth, the selective inhibitors of small GTPases such as Rho, Rac, or Cdc42 were applied to the cultured *Xenopus* embryonic spinal neurons that were stimulated with EFs (Fig. 5.1). This study revealed the regulation of growth cone turning by an EF. It also suggested that Rac and Cdc42 activities dominated cathodally, and Rho activity dominated anodally to guide growth cones toward the cathode. The study explored the temporally and spatially coordinated roles of Rho, Rac, Cdc42, and their signaling pathways in EF-directed growth cone migration [17].

In vitro studies showed that the electrical stimulation can enhance the chemotaxis induced by ECM guidance molecules. Naturally occurring molecules such as netrins, semaphorins, and neurotrophins can guide cones to grow toward or away from sources of the molecules [18]. Experiments were performed to examine the response of the growth cone of cultured *Xenopus* spinal neurons to the gradients of guidance molecules after the neuron was treated with electric stimulation. Electrical stimulation resulted in a significant change of growth cone turning of cultured *Xenopus* spinal neurons induced by gradients of guidance molecules. Electrical stimulation increased the response of netrin-1-induced attraction, while the stimulation converted the repulsion induced by myelin-associated glycoprotein (MAG) or myelin membrane fragments to attraction. This study suggested that the elevation of cytoplasmic Ca(2+) and cAMP mediated the process [19].

Both chemotaxis and electrotaxis play important roles in neural development. The growth cone navigation is guided by the integrated effects of multiple guidance cues. It was reported that the electrical activity is involved in growing axon path-finding in the developing nervous system [20, 21]. The electrical activity of neurons was crucial

5.1 Electric Field Directing Axonal Growth

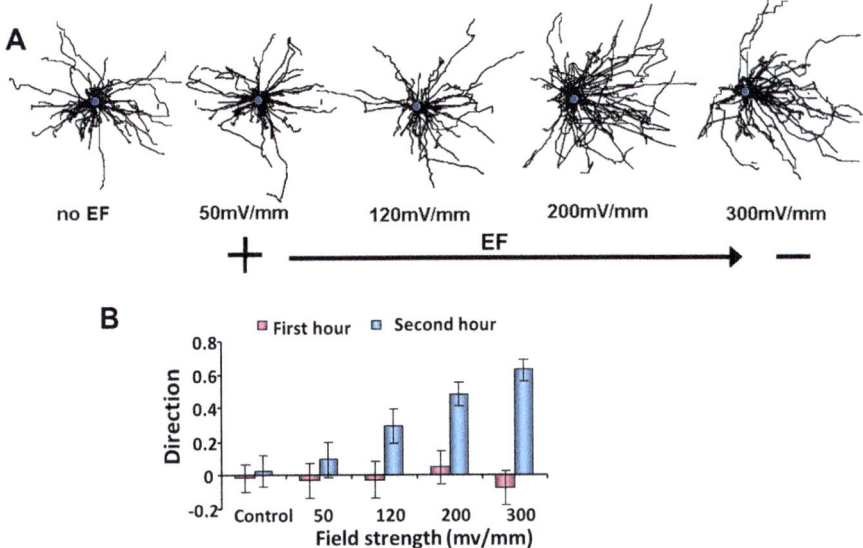

Fig. 5.1 (**a**) Net translocation of hippocampal neurons during 1-h migration with indicated EF strength. Migration paths determined by video monitor tracings. (**b**) Position of all cells at t = 0 min represented by origin position with migratory track of each cell at 60 min plotted as single line on graph. Applied EF strengths indicated in mV/mm. Figure reproduced from Yao et al. [44] with permission from John Wiley and Sons

for thalamic axonal growth toward the appropriate cortical region [1]. A number of research investigations showed that pharmacological and biological molecules can enhance, suppress, or switch the polarity of the guidance effect of the EF on neuron axons. The neurotransmitter acetylcholine gradient alone can direct the growth of *Xenopus* growth cones [22]. A muscarinic antagonist (atropine) can significantly enhance the cathodal axonal growth when the neuron was subjected to EF stimulation, while nicotinic antagonist (d-tubocurarine) can inhibit axonal cathodal growth in an applied EF [23]. The neurotrophin brain-derived neurotrophic factor (BDNF) enhanced the cathodal attraction of neuron axons in EFs. The endogenous cannabinoid anandamide can suppress the EF-induced cathodal growth of axons. Neurotrophin 3 (NT3) can reverse cathodal migration of axons to become anodal growth in an applied low EF [24, 25]. In vivo studies also demonstrated the impact of wound-induced EFs on nerve-growth regeneration. The wounded rat cornea can generate an endogenous EF immediately upon wounding. Wound-induced EFs with a cathode at the center of the wound can regulate the extent and direction of nerve-growth in vivo. Under the influence of EFs, sensory nerves close to the wound edge sprouted new processes and grew toward the wound center.

5.2 Electric Fields Directing Neuron Migration

Galvanotaxis is directional movement toward the cathode or the anode. Application of a DC EF produces galvanotaxis in a variety of cultured cells in vitro. Most cell types move toward the cathode; for example, bovine corneal epithelial cells (CECs) [26], bovine aortic vascular endothelial cells [27, 28], human retinal pigment epithelial cells [29], human keratinocytes [30], amphibian neural crest cells [31], C3H/10T1/2 mouse embryo fibroblasts [32], fish epidermal cells [33], and metastatic rat prostate cancer cells [34]. However, in some cases, cells move toward the anode; for example, human [35], rabbit corneal [36], and human umbilical vascular endothelial cells (HUVECs) [37]. It is also reported that human dermal melanocytes are insensitive to an external DC EF [38]. Species and cell subtype differences might affect galvanotaxis. It is reported that HUVECs move toward the anode [37], whereas bovine aortic vascular endothelial cells show a cathodal response [27].

Neuronal migration is composed of three schematic steps. First, the cell extends leading edges, preceded by filopodia and lamellipodia that explore the microenvironment. Growth cone extension proceeds following the integration of attractive and repulsive signals generated at the plasma membrane. The second step is nucleokinesis [39], which is directed by leading-edge extension and is dependent on the microtubule cytoskeleton. The third event in cell migration is retraction of the trailing process.

In the radial migration of the central nervous system (CNS), neuronal migration occurs along the glial processes. The neuronal leading processes elongate along the radial glial cells in a radio dimension of the neuroepithelium. Radial migration has been mostly studied in the cerebral cortex, hippocampus, and cerebellum and is referred to as "gliophilic migration" [40]. It accounts for the majority of neuronal migration events in the brain. In the CNS, neuronal migration following growth cone elongation is referred to as "neuro (no) philic" [41]. This type of neuronal migration occurs tangentially and superficially, close to the pial surface. The difference between growth cone elongation with and without cell migration depends on nucleokinesis following the leading process. Neuronal migration from subventricular zones (SVZs) in the lateral ganglionic eminences to the olfactory bulb (OB) (and possibly other forebrain areas) by following a glial tunnel is referred to as "chain migration." This type of neuronal migration extends through the tissue, mostly with a tangential orientation [41–43].

Injury or degenerative disease of CNS tissue causes the loss of neurons, and their recruitment is required to reestablish new synapses. Effective directional migration of endogenous and grafted neural cells can reconstruct functional connections and enhance the regeneration process. Neural crest cells of *Ambystoma mexicanum* and *Xenopus laevis* align perpendicular to an applied EF vector and migrate to the cathode [31]. In previous studies, the migration of both neurons and glial cells in EFs was investigated. Recent studies show that DC EFs can guide migration of dissociated neonatal hippocampal neurons and neurons from hippocampal explants to the cathode. At an early culture stage, hippocampal neurons showed bipolar morphology with short processes and dynamic migration. When neurons were exposed to an EF gradient, they migrated toward the cathode in the EF [44] (Fig. 5.1). Cathodal migration was enhanced by increasing the field strength. EFs steer the direction of neuron migration in three ways: by swapping over leading and trailing processes, by the growth cone turning and dragging the cell body to follow it, or by neurite branching with one branch dominating and leading neuronal cell body migration [14]. This indicates that EFs control the migration of neuronal cell bodies by guiding the leading growth cone as occurs in chemotaxis. The threshold effectiveness of EFs for directing hippocampal neuron migration lies between 50 and 100 mV/mm, although whether this varies with substrate has not been tested, and is comparable to physiologic thresholds of other cell types [45]. Previous work has shown that applied EFs had a stronger effect on the direction of cell migration, but whether migration speed was affected depended on cell type [18, 33, 45–48]. Neither EF-guided neuron migration nor chemorepulsion of neuron migration in a gradient of slit changed the migration speed [44, 46]. Both of these occurred during random neuronal migration and neuronal migration in an applied EF. The orientation of the leading neurite guided the neuron migration [49]. Neuron migration from hippocampal explants was also directed by an applied EF. When the cultured explants were subjected to an applied EF on their anodal-facing side, neurons migrated out and changed their direction to migrate more perpendicularly or turned cathodally in response to the EF stimulation. On the cathode-facing side of the explants, neurons migrated out and toward the cathode pole.

The guidance of the EF to the migration of glial cells was also reported. Oligodendrocyte precursor cells (OPCs) and Schwann cells myelinate the neuron axons in the central and peripheral nervous systems, respectively. Both cell types are involved in the myelination of regenerated axons in spinal cord injury (SCI). In the injured spinal cord, both OPCs and Schwann cells can regenerate and migrate to the lesion in order to myelinate axons [50–53]. The EF may effectively direct the migration of endogenous or grafted Schwann cells and OPCs to the lesion to enhance the remyelination process. When Schwann cells and OPCs isolated from a neonatal rat were stimulated with an applied EF, the cells migrated anodally. The cell migration behaved in a time- and voltage-dependent manner [54, 55].

5.3 EF-Guided Migration of Stem Cells and Stem Cell–Derived Neural Cells

Neurogenesis occurs during both the prenatal period and in the adult mammalian brain. Studies have shown that the adult hippocampus, olfactory bulb, ventricular epithelium, and subventricular zone possess various levels of neurogenesis [56–60]. Central nervous system disease such as brain trauma, stroke, degenerative disorders, and epilepsy stimulate the proliferation and differentiation of endogenous neural progenitor cells (NPCs) [61–66]. The neuron migration is critical to establishing new synapses in neurogenesis. The subventricular zone maintains the capability of generating NPCs in adult animals. These cells can move away from the SVZ and migrate along blood vessels toward the post-stroke striatum [67]. Similar to the SVZ, the hippocampus contains progenitor cells that can be activated after ischemia. The endogenous NPCs proliferate to generate hippocampal pyramidal neurons and migrate to the ischemia lesion of the hippocampus [68]. In adult mice, the neural precursors may differentiate into mature neurons in a layer- and region-specific manner after the corticothalamic neuron degeneration is induced. The new neurons from the precursors are involved in the establishment of the physiological corticothalamic connections [61]. The endogenous voltage gradients exist between damaged and normal brain tissue post-brain injury [69]. These studies indicate that the coordination of biological molecules and electrical signals established by injury direct the migration of neural precursors prior to the establishment of renewed and appropriate connections.

The migration of NPCs to the target location and the neural tissue lesion is essential for the development and repair of the CNS, respectively. The latent regenerative ability of neural progenitors could be activated to repair the injured neural tissue in adult mammals. Since the EF has demonstrated the ability to guide neural migration, it can be potentially used to direct transplanted neural stem cells (NSCs) or regenerated neurons to desired sites after brain injuries or neurodegeneration. Understanding the migration profile of NSCs can contribute to the improvement of therapeutic strategies for CNS repair. Neural stem cells show a clear cathodal migration in an applied electric field [70]. After NSCs were differentiated into OPCs, they showed the same cathodal migration in the EFs as the undifferentiated NSCs [71] (Fig. 5.2). The directedness and displacement of cathodal migration of the OPCs derived from NSCs increased significantly when the EF strength increased from 50 to 200 mV/mm. The NPCs of adult mouse brain migrated cathodally in the EFs. However, the cells lost their cathodal migration ability after the NPCs differentiated into astrocytes [72]. NPC replacement has been investigated as a therapy for neurodegenerative disorders such as Parkinson's disease (PD) [73, 74]. It was reported that a physiological level of EF stimulation regulated the directional migration of ventral midbrain NPCs, or NPCs(vm) [60], toward the cathode. The electrotactic response of cell migration was both time- and EF voltage-dependent. The study indicated that EF-directed migration of dopaminergic NPCs can potentially be used to treat PD.

Embryonic stem cells may promote endogenous neuronal regeneration and directly replace neurons, thereby improving physiological function recovery. Recent studies have reported the directional migration of induced pluripotent stem cells (iPSCs) and ESCs in an EF [75]. After human NSCs (hNSCs) derived from the human ESC (hESC) line H9 were stimulated with a small EF (16 mV/mm), the cells migrated to the cathode [76]. Reversal of the field polarity can reverse the cell migration to the new cathode pole. Increasing the stimulation time and voltage level enhanced the galvanotactic response. The cultured iPSCs generated from somatic cells of individual patients migrated anodally in an EF, while the expression of the human-induced pluripotent stem cell (hiPSC) markers, such as SSEA-4 and Oct-4, was not altered by an EF. The threshold to inducing EF-guided migration of cultured hiPSCs was 30 mV/mm. Spinal cord injury or disease can cause motor neuron loss that results in muscle paralysis. The neuron loss may be replaced by ESC-

5.3 EF-Guided Migration of Stem Cells and Stem Cell–Derived Neural Cells

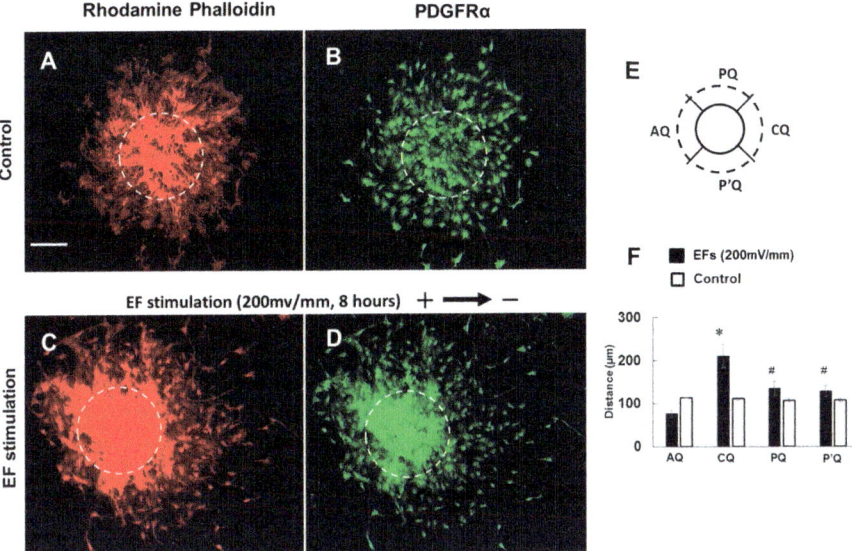

Fig. 5.2 Analysis of cell migration from oligospheres: (**a, b**) Symmetrical distribution of ARPC2+/+ NSC-OPCs around oligosphere without EF stimulation. (**c, d**) Asymmetrical distribution of NSC-OPCs around oligosphere exposed to EF for 8 h. Dashed cycle lines show core area of oligospheres. Scale bar: 100 μm. (**e**) Diagram showing division of quadrants around oligosphere. CQ is cathode-facing quadrant. AQ is anode-facing quadrant. PQ and P'Q are quadrants perpendicular field line of EFs. Solid circle line represents edge of oligosphere core. Dashed circle line represents frontier of migrated cells. (**f**) Analysis of cell migration distance from oligospheres. The *, $p < 0.01$, is compared with anode-facing quadrant and quadrants perpendicular to field line of EFs. The #, $p < 0.01$, is compared with anode-facing quadrant. Figure reproduced from Yao et al. [71] with permission from BioMed Central

derived motor neurons. The EF-directed ESC-derived motor neurons were reported recently. Both mouse ESC-derived NSCs and ESC-derived motor neurons migrated to the cathode in the EF [77].

Mesenchymal stem cells (MSCs) have demonstrated the capacity to promote regeneration of injured spinal cord when they were transplanted into injured spinal cord [78, 79]. MSCs are an attractive cell type in stem cell therapy because they can be easily obtained from somatic tissue such as bone marrow (BM), and the immune response of the host tissue to the MSCs is low in autologous transplantation. The EF-directed MSC migration was observed in in vitro studies. When human BM-MSCs were cultured in direct-current EFs of 10–600 mV/mm, the cells demonstrated an anodal migration in the EFs when the EF was above 25 mV/mm [80]. Cell migration reached the highest speed (42 ± 1 μm/h) when the applied EF was increased to 300 mV/mm. EF stimulation did not affect the cell senescence, phenotype, or osteogenic potential. An EF can be generated in wounded bone tissue and contribute to bone healing. The EF in the lesion may guide the migration of BM-MSCs to close the wound.

5.4 Oriented Cell Division in EFs

Cell division and migration play important roles in development, wound healing, and pathology. In the hippocampus of several species, including rat, mouse, rabbit, guinea pig, and cat, granule cells are generated postnatally [56–59]. Neurogenesis has been found in the cultures of an adult rat hippocampal slice culture study.

Evidence of neurogenesis was mostly derived from studies of the co-expression of the neuronal marker and BrdU. After injection of BrdU to 2-month-old adult rat brain, cell proliferation was investigated by means of immunofluorescence and confocal microscopy at several time points from 1 day to 11 months thereafter. BrdU-labeled neurons remained stable in number and in their relative position in the granule cell layer over at least 11 months. These results suggest that the addition of new neurons is not transient and that their final number and localization are determined early [81]. Neurogenesis was also found in an adult rat hippocampal slice culture study. After a four-week culture of hippocampal slices in vitro, confocal images revealed that some of the BrdU-labeled nuclei were co-expressed with NeuN in the GCL. These cells were not immunoreactive with anti-glial fibrillary acidic protein (GFAP). The co-expression of BrdU and NeuN was examined three-dimensionally under higher magnification. In the layer overlaying the GCL, there were some BrdU-labeled nuclei surrounded by GFAP-positive processes. These cells were not immunolabeled by anti-NeuN [82].

In the rat, neurogenesis continues well into adulthood. The newborn granule cells are capable of extending axonal projections along the mossy fiber tract to their natural target area, the hippocampal CA3 region [60], and exhibit all ultrastructural features associated with neurons [83].

The axes of cell division have a major morphogenetic impact. In the developing mouse central nervous system (CNS), the axis of cleavage of division determines the fate of the daughter cells [84]. Oriented division and directional migration are essential for correctly locating postmitotic neurons in the developing CNS.

Both intrinsic and extrinsic factors influence cell division and cause a degree of order [85]. The physiological role of EF-induced regulation of the cell division plane has been shown in in vitro experiments and also indicated in in vivo studies. Both primitive streak formation and neurulation and maturation of the neural tube are associated with the transient appearance of spatially restricted small DC EFs [86, 87]. Disrupting the endogenous EFs in the neural plate and neural tube of amphibia can cause specific abnormalities during CNS development [87, 88].

During neurulation in zebra fish, rats, and chicks, neuroepithelial cells divide either along a rostrocaudally oriented spindle axis or at 90° to this, mediolaterally [85, 89, 90]. Two orthogonal EFs with rostrocaudal and mediolateral vectors have been observed [87] when measuring spatial variations in the transepidermal potential difference across the developing amphibian neural plate. Electrical activity is also involved in the control of Schwann cell division. Saltatory conduction between nodes involves extracellular current flow and will establish pulsed, extracellular EFs oriented parallel to the nerve, which could influence Schwann cell mitotic spindle orientation. It is reported that when electric activity was inhibited by tetrodotoxin, cell division was inhibited [91].

During corneal epithelial wound healing, DNA synthesis and mitotic activity near a wound edge increased markedly during the first 12 h. [92]. Wounds in rat corneas generate endogenous EFs in the plane of the epithelial sheet because the transcorneal potential difference (TCPD) collapses at the wound edge. In vitro experiments have shown that a wound-induced EF controlled the orientation of cell division; most epithelial cells divided with a cleavage plane parallel to the wound edge and perpendicular to the EF vector [93].

The observation of external applied EFs to corneal epithelial cells also proved the role of electric activity in controlling cell division direction. When cultured human CECs were exposed to a direct-current EF of physiological magnitude, cells were divided with a cleavage plane perpendicular to the EF vector [94]. Hippocampal cells were divided in EFs, and the cell division plane showed a preferred angle in the EFs (Fig. 5.3). Most cell division planes were perpendicular to the field line. This study suggests that the EF is an important external factor in controlling hippocampal neuronal cell division. There are several scenarios in which hippocampal neurons may be exposed to physiological DC EFs at times when cell division is taking place. This makes it highly likely that the axis of neuronal and glial cell division will be regulated by an EF in specific regions of the CNS.

5.5 Regulation of EF-directed Neuronal Migration

5.5.1 Cell Polarization in EFs

Cell polarization is a cellular response to environmental cues. Directional sensing, cell polarization, and directional migration are a series of associated events in directional cell migration. During this process, cells are dynamically polarized in response to an extracellular gradient of signals. Migrating fibroblasts exhibit a polarized cell shape with a membrane ruffling area and filopodia. In migrating fibroblasts during wound-induced healing, the characterized morphology of polarization is membrane ruffling and filopodia at the front edge. Stimulated with EFs, Chinese hamster ovary (CHO) cells elongated and polarized with distinct leading-edge lamellipodia facing cathodally [95], while endothelial cells oriented perpendicular to the EFs [37]. Neurons are highly polarized cells and contain two distinct types of processes, an axon and dendrites, which allow them to receive, process, and transmit information. In an axon, the microtubules are uniformly oriented with their plus ends distal to the cell body, whereas in dendrites, the MTs plus ends are non-uniformly oriented [96]. Not only shown in the morphology change, cell polarization also requires asymmetric distributions of membrane receptors, intracellular signaling molecules, and cytoskeletons, as well as directed membrane trafficking.

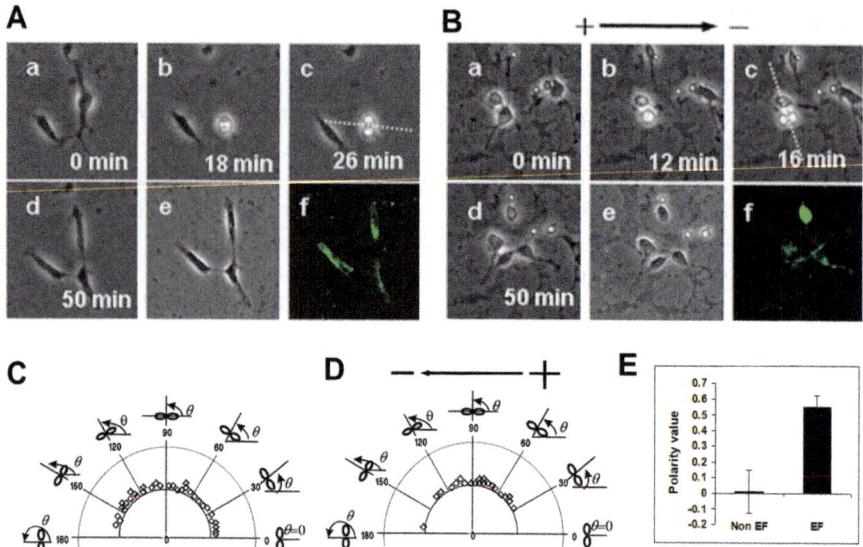

Fig. 5.3 Cell division in EFs, recorded using time-lapse imaging: (**a**) Cell dividing without EF application. (**b**) Cell dividing in applied EF of 300 mV/mm. White dashed lines drawn in images A(c) and B(c) highlight cleavage planes of neuronal cell division. At conclusion of experiments, cells fixed with 4% paraformaldehyde. Images A(e) and B(e) show cells after fixation. Daughter cells following division labeled with neuron-specific anti-mitogen-activated protein (MAP)-2 antibody. (**c–e**) Analysis of EF effect on cell cleavage plane. Angles of cleavage plane of rat hippocampal neurons expressed as polar diagram, where each symbol represents one cell. Cell division followed with time-lapse recording. (**c**) Angles of cleavage plane of 28 dividing cells cultured without EFs. (**d**) Polarized orientation of cleavage plane of 21 cells dividing in small EF (50–300 mV/mm). (**e**) Mean oriented division index of cleavage plane. Figure reproduced from Yao et al. [49] with permission from John Wiley and Sons

In chemotaxis, a gradient of extracellular molecules stimulates activation of more receptors on one side of the cell than the other. In electrotaxis, an increase in density of membrane receptors on the cathode-facing side results in asymmetry of the signaling [97]. Several growth factor receptors may act together and use parallel signaling pathways to transduce the effects of an EF. Studies have shown that in serum-free medium, bovine corneal epithelial cells lose all cathodal directionality in migration with respect to the EF vector. The addition of epidermal growth factor (EGF), basic fibroblast growth factor (bFGF), or transforming growth factor (TGF)-β1 to a serum-free culture medium restored directed migration partially and was dose dependent. However, when combinations of the three growth factors were added to a serum-free medium, cathodal migration was fully restored. Cathodal accumulation of EGF receptors and of membrane surface area of epithelial cells were identical quantitatively [98] due to the membrane area increase cathodally because of membrane folding.

The establishment of cell polarity means that the molecular processes at the front and the back of a moving cell are different, and this is mediated by a set of interlinked positive-feedback loops involving Rho-family GTPases, phosphoinositide

3-kinases (Pi3ks) [99]. Studies of neutrophil and *Dictyostelium* chemotaxis showed that Pi3ks are key mediators during cell polarization and cell migration. The pleckstrin homology (PH) domain containing proteins, such as Akt or protein kinase B (PKB), specifically interact with Pi3k products, for example, phosphatidylinositol 3,4-bisphosphate (PIP2) and phosphatidylinositol 3,4,5-trisphosphate (PIP3). A green fluorescent protein (GFP)-fused PH domain containing protein enables us to visualize the activity of Pi3k spatiotemporally [100]. When neutrophil-like cells from the HL-60 cell line are exposed to a chemoattractant gradient delivered from a micropipette, PH-Akt-GFP rapidly and transiently translocated toward the chemoattractant [100]. Rho-family GTPases, including Cdc42, Rac1, and RhoA regulate the cytoskeleton, cell adhesion [101, 102], and cell polarization in several cell types, such as fibroblasts [103], astrocytes [104, 105], epithelial cells [106], and neuronal cells [107]. The interaction of Rho-family GTPases and Pi3kase regulates cell polarity during cell migration. It is suggested that a positive-feedback loop exists during cell polarization (i.e., PIP3-GEFs-Rho GTPases-Pi3k-PIP3) [96]. An inhibitor of Pi3k or RhoA can decrease or even abolish the migration direction and orientation of several cell types in EFs [37, 95].

The chemical gradient or electric signals activate membrane receptors and downstream intracellular signaling elements, and this leads to asymmetrical distribution of the cytoskeleton. The exposure of corneal epithelial cells to physiological electric fields showed the activation of the extracellular signal-regulated kinase (ERK) and accumulation of F-actin at the leading, cathodal-facing side of the cell [97]. During wound-induced healing, concomitant with cell morphology polarization, the migrating fibroblasts also reoriented their microtubule-organizing center (MTOC) and the Golgi apparatus in the direction of cell migration. Polarization of the Golgi apparatus can help the anterograde supply of membrane components to the leading edge for membrane protrusion. Signals from the Golgi matrix allow reorientation of the Golgi toward the direction of movement [108, 109]. Microtubules form a perinuclear cage-like structure converging into the centrosome and projecting into the leading process from the centrosome. Golgi and centrosome polarization are important elements in directional cell migration. Scratch-wounding monolayer cultures of fibroblasts, astrocytes, and endothelial cells generate reproducible and consistent cues to induce Golgi polarization [103–105, 110, 111]. In random neuronal migration, the leading process leads the way of neurons, and the Golgi-centrosome is located at the base of the neurite of migrating neurons [112]. The translocation of the centrosome and the Golgi occurred during the reversal of migration in inhibitory neurons.

CHO cells exhibited a polarized orientation of the Golgi apparatus to the cathode-facing side when subjected to an applied EF. The Golgi polarization of CHO cells in an EF can be decreased with a Pi3k inhibitor [95]. The cooperation among the actin and MTs and the regulation of intracellular molecules to the cytoskeleton are required. Golgi and the centrosome showed a high degree of polarization in neurons subjected to EF stimulation. Polarization of the Golgi and centrosome was in the same direction as the EF-guided neuronal migration. The Golgi were motile in neurons, and the movement of Golgi from the trailing process to the leading process was observed in this study.

5.5.2 Calcium Signals Growth Cone Navigation and Neuron Migration in Applied EF

EF stimulation exerts an electrophoretic force on charged proteins and lipids in the plasma membrane and therefore results in a redistribution of membrane components [113–116]. Calcium signaling may play a critical role in EF-directed cell migration. Induced by DC electric fields, cell membrane receptors accumulate on the cathode- or anode-facing side of the growth cone and therefore activate signaling cascades. The redistribution of ion channels may form regional clusters and increase local fluxes of ions. Ca^{2+} influx through the voltage-gated Ca^{2+} channels (VGCCs) contributes to the growth cone cathodal turning [117]. Calcium entry causes the isolation of the actin meshwork of lamella facing the cathode and leads to the protrusion of the lamellipodia toward the cathode [27, 97, 118]. The transition area between the leading lamella facing the cathode and the cell body will also experience EF-induced local calcium entry, which activates the actomyosin contractile system, thereby pulling the cell forward. The increased tension will cause retraction on the anode-facing side [119]. Calcium elevation can also cause the increase of the intracellular cAMP level via adenylate cyclase. In turn, cAMP activates protein kinase A (PKA) that signals small GTPases and therefore regulates the dynamics of filamentous actin and microtubules that steer the growth cone migration [120]. It was reported that the voltage-gated sodium (Na) channel (VGNC) is also involved in the electrotaxis. Electrotaxis can be suppressed by tetrodotoxin, which is a VGNC blocker, while veratridine can prolong the Na^+ channel opening and enhance electrotaxis [121].

5.5.3 Cell Membrane Receptors and Intracellular Signaling Pathways

The mechanism of EF-directed axonal growth and cell migration is not fully understood. However, studies have revealed that the polarization of cells, activation of ion channels, and intracellular signaling pathways are involved in the regulation of EF-guided migration. In chemotaxis, the polarized cells migrate dynamically in response to extracellular signaling gradients. When the cells are subjected to EF stimulation, a uniform current flows over the top and bottom of cells. Because the cell plasma membrane poses a large resistance to current flow, ionic currents driven through the medium outside the cell by the imposed EF will be forced to flow mainly around the cell. Plasma membranes pose a large electrical resistance to current flow. Current density increases along the sides of the cells parallel to the EF vectors, and the regional EF may increase by about 50% [122]. Such fields impose a local membrane potential perturbation (depolarizing and hyperpolarizing the cathode and anode-facing membranes, respectively) and also generate a lateral voltage gradient along the upper and lower membrane surfaces, which in turn imposes a redistribution of charged membrane components through the process of electroosmosis [47, 114, 122].

Integrins play an important role in EF-guided cell migration because they mediate cell adhesion and migration. The function of integrin α3 in the migration of neural stem/progenitor cells (NSPCs) and glioma cells was reported. The integrin α3 regulated extracellular matrix-dependent NSPC migration [123]. Overexpression of integrin a3 enhanced glioma cell migration and invasion, while the downregulation of integrin α3 inhibited the cell invasion [124]. It was reported that integrin α2β1 was involved in regulating the directional migration of a ligament fibroblast in an EF. The blocking of integrin α2β1 with its antibody significantly attenuated EF-directed cell migration while the cells remained motile [125]. The role of integrin β4 in the directional migration of normal human keratinocytes (NHKs) in EFs was also revealed in a previous study [126]. The integrin β4 is needed in the EF-directed cathodal migration of NHKs when the epidermal growth factor is absent. NHKs lost their directional migration in EFs in the absence of EGF when the cells expressed either a ligand-binding-defective beta4 (β4 + AD) or beta4 with a truncated cytoplasmic tail (β + CT). This study suggested that the beta4 integrin ligand-binding and activation of the EGF receptor can synergistically activate the Rac-dependent signaling pathway. Either pathway can independently regulate the directional migration of keratinocyte in response to EF stimulation.

A number of signaling pathways including Pi3k, phosphatase and tensin homolog (PTEN), and small Rho GTPases were found to regulate the EF-guided neural cell migration. The role of Pi3k in cell migration is involved in determining migratory directedness and the motility of cell movement undergoing chemotaxis. The role of Pi3k in cell migration directedness is mainly derived from the evidence of single-cell organisms or neutrophil migration. Neutrophil lacking Pi3k showed the defect of migration directedness in chemotaxis [127]. In vitro experiments on brain slices showed that the Pi3k inhibitor LY294002 completely suppressed the tangential migration of medial ganglionic eminence (MGE) neurons to the cortex in the co-culture assay [128]. When an SVZ explant culture was exposed to LY294002 in vitro, the neuronal migration was significantly reduced.

Many types of mammalian cells showed directional migration in EFs. Pi3k reduced epithelial cell migration directedness in EFs. When epithelial cells were treated with LY294002, they lost cathodal directedness. Genetic disruption of Pi3kγ abolished directed movements of the wounded epithelium layer in response to EF stimulation, while deletion of PTEN enhanced the electrostatic responses and directional cell migration [129]. The cathodal migration of hippocampal neuron in EFs was decreased after neurons were treated with LY294002 [44]. This Pi3k inhibitor also influenced neurite orientation of Xenopus spinal neurons in EFs. Xenopus spinal neurites lost cathodal turning when exposed to LY294002 [17]. Because the neuron soma movement is guided by the growth cone and the orientation of the leading process in EFs, the defect of leading-process cathodal orientation can cause the loss of neuronal migration directedness.

RhoA and its effectors ROCK1 and ROCK2 are key modulators of cytoskeletal dynamics in cell adhesion and cell migration [130–132]. Inhibition of ROCK1/2 abolishes precerebellar neuron nuclear migration toward a netrin-1 source in collagen assays [133]. Additionally, the inhibition of ROCK function can reduce neurite retraction [134, 135]. The role of ROCK in neuronal migration has shown that during tangential migration of PCN in response to netrin-1, a ROCK inhibitor can cause axon extension and block neuronal migration [133]. The EF-guided directional migration of hiPSCs was inhibited by ROCK inhibitor Y27632, while Y27632 did not significantly affect the directional migration of human NSCs in EFs [76]. The neurite orientation and neuronal migration directedness in EFs were decreased significantly by pre-treating with Y27632. Results suggest that ROCK is a crucial molecule in determining neuronal migratory motility and directedness [44]. Neuronal stem/progenitor cells (NSPCs) migrated toward the cathode in EFs. The activation of N-methyl-D-aspartate receptors (NMDARs) led to an increased physical association of Rho GTPase Rac1 signals to the membrane NMDARs and the intracellular actin cytoskeleton, thereby regulating EF-guided NSPC migration [136].

Akt is involved in the control of directional cell migration and the sensing of chemoattractant gradients. GFP-Akt robustly and constantly localized at the leading edge of randomly migrating HT1080 cells [137]. In neurons, Akt is upstream of GSK-3βin that determines neuronal polarity [138]. During chemotaxis, Pi3k plays a function in cell polarization via facilitation of Akt phosphorylation [139]. *Dictyostelium* and mammalian Akt/PKB PH domain fusions, CRAC and the protein PhdA differentially translocate to the plasma membrane in response to global stimulation by a chemoattractant and to the leading edge in response to directional chemoattractant stimulation [140]. An EF is another type of cue directing cell migration, and it also offers an easy and useful laboratory tool to analyze the cellular and molecular mechanisms of directional cell behavior. When CHO cells were stimulated with EFs, Akt was activated asymmetrically on the cathode side of the cell, which is consistent with the cell migration direction [95]. Consistent with the observation in fibroblasts, when exposed to EFs, Phospho-Akt in neurons was activated asymmetrically on the cathode-facing side [44].

It was reported that the mitogen-activated protein kinase (MAPK) ERK1/2 was activated in migration cells such as epithelial cells, fibroblasts, and vascular smooth muscle cells. The extracellular signal-regulated kinase cascade regulates both cell motility and cell growth. EF stimulation activated and polarized the ERK1/2 of mammalian epithelial cells and was involved in the directional migration of cells in the EF [97]. When corneal epithelial cells were subjected to EF stimulation, the ERK was activated and F-actin accumulated at the leading edge of the cathode-facing side of the cell. EF-polarized membrane lipid domains and EGF receptors caused asymmetric signaling that activates MAPK and therefore induced directional cell migration.

In a recent study, the signaling pathways of EF-guided cell migration was systemically investigated using next-generation RNA sequencing [54] (Fig. 5.4). The differentially expressed genes of Schwann cells subjected to EF stimulation revealed the pathways that regulate cell migration, such as the actin cytoskeleton and focal adhesion pathways. These pathways regulate the function of intracellular cytoskeletons, which are critical for cell motility. This study found that EF significantly altered the expression of a few integrin transmembrane receptor genes. In comparison with the non-EF-treated cells, the expression of integrin α3 in EF-stimulated Schwann cells was up-regulated, while integrin α4, integrin α11, integrin β4, and

5.6 Electrical Activity in CNS Development and Regenerating Tissues

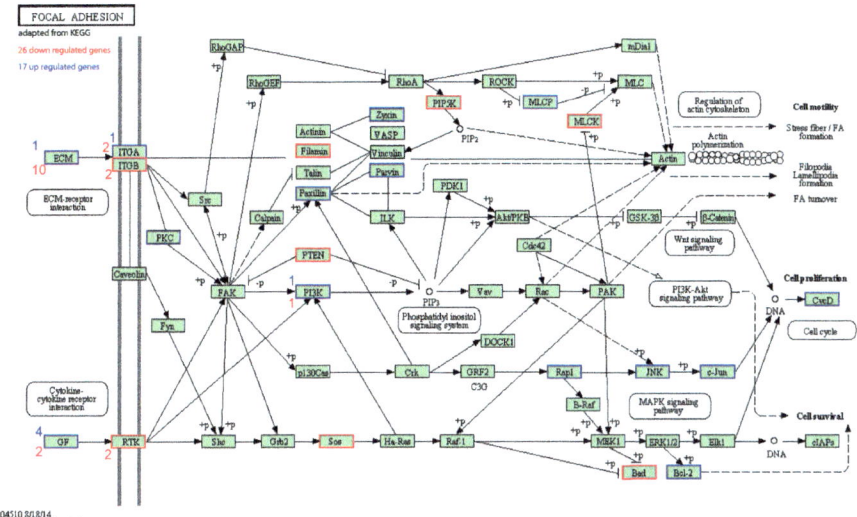

Fig. 5.4 Signaling pathway of focal adhesion. Red-color-labeled boxes show down-regulated genes. Blue-color-labeled boxes indicate up-regulated genes. Purple-color-labeled boxes contain both up- and down-regulated genes. Figure reproduced from Yao et al. [54] with permission from John Wiley and Sons

integrin β8 were down-regulated. The expression of the PIK3R3 gene was down-regulated, while the expression of PIK3R5 and PTEN genes was up-regulated in EF-treated Schwann cells [54]. This finding confirmed that PTEN down-regulation is required for EF-guided cell migration. The molecules on the ITGB-PKC-FAK-Paxillin-Parvin-Actin pathway regulated actin polymerization. Parvin is an adapter protein that interacts with other focal adhesion proteins such as Paxillin and ILK, leading to focal adhesion stabilization. The expression of PKC, Paxillin, and Parvin was up-regulated in cells subjected to EF stimulation.

5.6 Electrical Activity in CNS Development and Regenerating Tissues

Endogenous EFs are widespread in developing and regenerating tissues. The occurrence of endogenous currents has been found in many developing system [86, 141–144]. The extracellular EFs are detected using glass microelectrodes, and the associated currents are measured using a vibrating probe. During embryonic development, endogenous EFs have been detected at the neural plate and neural fold stages. The EF vectors are aligned with the major embryonic axes [172]. The size, location, and developmental timing of EFs are appropriate to influence cell behavior during nervous system development. In the neural tube, EFs are larger across the developmentally crucial floor plate than they are laterally [145]. When endogenous EFs were disrupted selectively [88, 146–148], a specific array of developmental defects, including failure of neural tube closure and the absence of some CNS structures, limb buds, or tail, can be observed.

A previous study observed the response of cultured neural crest cells of *Ambystoma mexicanum* and *Xenopus laevis* to DC EFs. When cultured neural crest cells were subjected to DC EFs, the cells aligned perpendicular to the EF vector and migrated to the cathode. Results support the view that EFs could be one of the embryonic environmental factors in neural crest pathways and provides factors that support and perhaps guide the movements of neural crest cells and of motile cells in general [173].

An endogenous EF generates immediately upon wounding and penetrates about 1 mm back from the wound edge. Wound-generated endogenous electric currents and the endogenous electric fields were detected in the wound-healing process [93, 149, 150]. EFs can direct the migration of cells toward the cathode, which is at the center of the wound, thereby promoting wound healing [93, 151]. The proliferation and migration of epithelial cells decrease the size of the wound. Sensory nerves sprouted new processes and grew toward the wound edge in the area close to the wound [152].

An electric current was also detected in the wounded rat hippocampus using a vibrating probe. The outward currents before wounding were reverted to the inward current immediately after wounding. The inward current lessened with time. This electric current change was also observed in mouse and human skin wounds [153]. The steady EF is established by the ion current in a wounded hippocampus because of mechanical injury to brain tissue. The wound-induced EFs could attract neurons or transplanted neural cells to migrate into these wounded areas to establish functional connections.

5.7 Electric Fields Enhance Nerve and Spinal Cord Regeneration In Vivo

Electrical stimulation for peripheral nerve regeneration has been studied in previous research. When a crushed rat sciatic nerve was stimulated with EF, the nerve regrowth and blood supply were increased [154]. The alternating current stimulation significantly promoted reinnervation for the transected femoral nerve or crushed sciatic nerve [155, 156]. The combinatorial application of a biomaterial neural conduit and electrical stimulation has been studied to repair the nerve defect. A 15-mm rat sciatic nerve defect was repaired using a conductive scaffold. Intermittent electrical stimulation (3 V, 20 Hz) significantly promoted nerve regeneration and axon remyelination [157].

Axonal regeneration after injury or disease is the major challenge in the central nervous system. Studies have demonstrated that EF stimulation can promote axonal regeneration in wounded spinal cord cathodally [158–160]. The application of a steady electric current (approximately 10 µA) to a completely severed spinal cord of the larval lamprey *Petromyzon marinus* enhanced axonal regeneration of the severed giant reticulospinal axons [158]. EFs also have been applied to partially severed spinal cords of adult guinea pigs using an implanted battery and electrodes [161]. The applied electric currents by implanted batteries can enhance axonal regeneration in the severed spinal cord. The rat spinal cord following contusion injuries was stimulated with an EF, which promoted axonal growth of the descending axonal tracts [160].

EF stimulation has been investigated to treat spinal cord injury in clinical trials. In one study, 14 patients with complete motor and sensory SCI were stimulated with oscillating field stimulation. The patients showed improvement in light touch, pinprick, and motor scores [162]. This study demonstrated the safety and reliability of oscillating field stimulation to patients post-SCI.

5.8 Potential of Applying Electric Fields to Guide Cell Migration in Neurogenesis

Electrical stimulation and recording techniques are currently playing important roles in diagnostic, and therapeutic strategies of nervous system disease. However, the highly complex CNS structure and function is a great challenge to effectively applying electric fields to promote neural regeneration in both the peripheral and central nervous system. In vivo studies have shown that both endogenous and applied EFs can guide and promote axonal regeneration. Neuron regeneration is required to replace neuron loss in injury or degenerative diseases of CNS tissues to establish new functional connections. Therefore, directional migration of endogenous and transplanted neurons is critical in neural repair and the regeneration process.

A number of cell types participate in the neural regeneration process post-SCI. Both proteolytic and non-proteolytic mechanisms are involved in the neural cell migration in ECM. In protease-independent migration, the deformation of the cell body and nucleus allows the cells to fit themselves into the pre-existing space [163, 164]. The cells may be able to reorient the migration route and bypass the obstacle [165]. The migrating cells may also migrate through the extracellular matrix by degrading the pericellular matrix. The cathepsins, serine proteases, and matrix metalloproteinases (MMPs) contribute to degradation of the ECM and proteolytic migration [163, 166, 167]. The naturally occurring electric current flow from subventricular zone to the olfactory bulb in an adult mouse brain was detected. The electric potential between SVZ and OB is about 3 mV/mm. In one study, the endogenous electric currents may contribute to the rostral neuronal migration. In one study, EF was applied to the cultured brain slices. The applied EF guided the migration of neuroblasts from the SVZ [168].

Electrical stimulation such as deep brain stimulation and epidural stimulation has been investigated for CNS disease therapy. These studies have demonstrated the safety and effectiveness of electrical stimulation in treating CNS disease. In the treatment of CNS malfunction, such as Parkinson's disease, traumatic brain injuries, paralysis, multiple sclerosis, and depression, deep brain stimulation significantly affected neural recovery [169–171].

References

1. Catalano SM, Shatz CJ. Activity-dependent cortical target selection by thalamic axons. Science. 1998;281(5376):559–62.
2. Dantzker JL, Callaway EM. The development of local, layer-specific visual cortical axons in the absence of extrinsic influences and intrinsic activity. J Neurosci. 1998;18(11):4145–54.
3. Tessier-Lavigne M, Goodman CS. The molecular biology of axon guidance. Science. 1996;274(5290):1123–33.
4. Mueller BK. Growth cone guidance: first steps towards a deeper understanding. Annu Rev Neurosci. 1999;22:351–88.
5. Bentley D, O'Connor TP. Cytoskeletal events in growth cone steering. Curr Opin Neurobiol. 1994;4(1):43–8.
6. Tanaka E, Sabry J. Making the connection: cytoskeletal rearrangements during growth cone guidance. Cell. 1995;83(2):171–6.
7. Song H, Poo M. The cell biology of neuronal navigation. Nat Cell Biol. 2001;3(3):E81–8.
8. Jaffe LF, Poo MM. Neurites grow faster towards the cathode than the anode in a steady field. J Exp Zool. 1979;209(1):115–28.
9. Hinkle L, McCaig CD, Robinson KR. The direction of growth of differentiating neurones and myoblasts from frog embryos in an applied electric field. J Physiol. 1981;314:121–35.
10. Patel N, Poo MM. Orientation of neurite growth by extracellular electric fields. J Neurosci. 1982;2(4):483–96.
11. McCaig CD. Dynamic aspects of amphibian neurite growth and the effects of an applied electric field. J Physiol. 1986;375:55–69.
12. Li S, Li H, Wang Z. Orientation of spiral ganglion neurite extension in electrical fields of charge-balanced biphasic pulses and direct current in vitro. Hear Res. 2010;267(1–2):111–8.
13. McCaig CD, Rajnicek AM, Song B, Zhao M. Has electrical growth cone guidance found its potential? Trends Neurosci. 2002;25(7):354–9.
14. McCaig CD. Spinal neurite reabsorption and regrowth in vitro depend on the polarity of an applied electric field. Development. 1987;100(1):31–41.
15. Patel NB, Poo MM. Perturbation of the direction of neurite growth by pulsed and focal electric fields. J Neurosci. 1984;4(12):2939–47.
16. Park JS, Park K, Moon HT, Woo DG, Yang HN, Park KH. Electrical pulsed stimulation of surfaces homogeneously coated with gold nanoparticles to induce neurite outgrowth of PC12 cells. Langmuir. 2009;25(1):451–7.
17. Rajnicek AM, Foubister LE, McCaig CD. Temporally and spatially coordinated roles for Rho, Rac, Cdc42 and their effectors in growth cone guidance by a physiological electric field. J Cell Sci. 2006;119.(Pt 9:1723–35.
18. Caroni P. Driving the growth cone. Science. 1998;281(5382):1465–6.
19. Ming G, Henley J, Tessier-Lavigne M, Song H, Poo M. Electrical activity modulates growth cone guidance by diffusible factors. Neuron. 2001;29(2):441–52.
20. Shatz CJ, Stryker MP. Prenatal tetrodotoxin infusion blocks segregation of retinogeniculate afferents. Science. 1988;242(4875):87–9.
21. Goodman CS, Shatz CJ. Developmental mechanisms that generate precise patterns of neuronal connectivity. Cell. 1993;72(Suppl):77–98.
22. Zheng JQ, Felder M, Connor JA, Poo MM. Turning of nerve growth cones induced by neurotransmitters. Nature. 1994;368(6467):140–4.
23. Erskine L, McCaig CD. Growth cone neurotransmitter receptor activation modulates electric field-guided nerve growth. Dev Biol. 1995;171(2):330–9.
24. McCaig CD, Sangster L, Stewart R. Neurotrophins enhance electric field-directed growth cone guidance and directed nerve branching. Dev Dynam. 2000;217(3):299–308.
25. Berghuis P, Rajnicek AM, Morozov YM, Ross RA, Mulder J, Urban GM, et al. Hardwiring the brain: endocannabinoids shape neuronal connectivity. Science. 2007;316(5828):1212–6.

References

26. Zhao M, Agius-Fernandez A, Forrester JV, McCaig CD. Directed migration of corneal epithelial sheets in physiological electric fields. Invest Ophthalmol Vis Sci. 1996;37(13):2548–58.
27. Li X, Kolega J. Effects of direct current electric fields on cell migration and actin filament distribution in bovine vascular endothelial cells. J Vasc Res. 2002;39(5):391–404.
28. Bai H, McCaig CD, Forrester JV, Zhao M. DC electric fields induce distinct preangiogenic responses in microvascular and macrovascular cells. Arterioscler Thromb Vasc Biol. 2004;24(7):1234–9.
29. Sulik GL, Soong HK, Chang PC, Parkinson WC, Elner SG, Elner VM. Effects of steady electric fields on human retinal pigment epithelial cell orientation and migration in culture. Acta Ophthalmol. 1992;70(1):115–22.
30. Sheridan DM, Isseroff RR, Nuccitelli R. Imposition of a physiologic DC electric field alters the migratory response of human keratinocytes on extracellular matrix molecules. J Invest Dermatol. 1996;106(4):642–6.
31. Cooper MS, Keller RE. Perpendicular orientation and directional migration of amphibian neural crest cells in dc electrical fields. Proc Natl Acad Sci U S A. 1984;81(1):160–4.
32. Onuma EK, Hui SW. A calcium requirement for electric field-induced cell shape changes and preferential orientation. Cell Calcium. 1985;6(3):281–92.
33. Cooper MS, Schliwa M. Motility of cultured fish epidermal cells in the presence and absence of direct current electric fields. J Cell Biol. 1986;102(4):1384–99.
34. Djamgoz MBA, Mycielska M, Madeja Z, Fraser SP, Korohoda W. Directional movement of rat prostate cancer cells in direct-current electric field: involvement of voltagegated Na+ channel activity. J Cell Sci. 2001;114.(Pt 14:2697–705.
35. Rapp B, de Boisfleury-Chevance A, Gruler H. Galvanotaxis of human granulocytes. Dose-response curve. Eur Biophys J. 1988;16(5):313–9.
36. Chang PC, Sulik GI, Soong HK, Parkinson WC. Galvanotropic and galvanotaxic responses of corneal endothelial cells. J Formosan Med Assoc. 1996;95(8):623–7.
37. Zhao M, Bai H, Wang E, Forrester JV, McCaig CD. Electrical stimulation directly induces pre-angiogenic responses in vascular endothelial cells by signaling through VEGF receptors. J Cell Sci. 2004;117.(Pt 3:397–405.
38. Grahn JC, Reilly DA, Nuccitelli RL, Isseroff RR. Melanocytes do not migrate directionally in physiological DC electric fields. Wound Repair Regen. 2003;11(1):64–70.
39. Morris NR, Efimov VP, Xiang X. Nuclear migration, nucleokinesis and lissencephaly. Trends Cell Biol. 1998;8(12):467–70.
40. Rakic P. Radial unit hypothesis of neocortical expansion. Novartis Found Symp. 2000;228:30–42. discussion -52
41. Luskin MB. Restricted proliferation and migration of postnatally generated neurons derived from the forebrain subventricular zone. Neuron. 1993;11(1):173–89.
42. Wichterle H, Garcia-Verdugo JM, Alvarez-Buylla A. Direct evidence for homotypic, glia-independent neuronal migration. Neuron. 1997;18(5):779–91.
43. Law AK, Pencea V, Buck CR, Luskin MB. Neurogenesis and neuronal migration in the neonatal rat forebrain anterior subventricular zone do not require GFAP-positive astrocytes. Dev Biol. 1999;216(2):622–34.
44. Yao L, Shanley L, McCaig C, Zhao M. Small applied electric fields guide migration of hippocampal neurons. J Cell Physiol. 2008;216(2):527–35.
45. Nuccitelli R. Ionic currents in morphogenesis. Experientia. 1988;44(8):657–66.
46. Ward M, McCann C, DeWulf M, Wu JY, Rao Y. Distinguishing between directional guidance and motility regulation in neuronal migration. J Neurosci. 2003;23(12):5170–7.
47. Nishimura KY, Isseroff RR, Nuccitelli R. Human keratinocytes migrate to the negative pole in direct current electric fields comparable to those measured in mammalian wounds. J Cell Sci. 1996;109. (Pt 1:199–207.
48. Farboud B, Nuccitelli R, Schwab IR, Isseroff RR. DC electric fields induce rapid directional migration in cultured human corneal epithelial cells. Exp Eye Res. 2000;70(5):667–73.
49. Yao L, McCaig CD, Zhao M. Electrical signals polarize neuronal organelles, direct neuron migration, and orient cell division. Hippocampus. 2009;19(9):855–68.

50. Almad A, Sahinkaya FR, McTigue DM. Oligodendrocyte fate after spinal cord injury. Neurotherapeutics. 2011;8(2):262–73.
51. Wang S, Sdrulla AD, diSibio G, Bush G, Nofziger D, Hicks C, et al. Notch receptor activation inhibits oligodendrocyte differentiation. Neuron. 1998;21(1):63–75.
52. Guest JD, Hiester ED, Bunge RP. Demyelination and Schwann cell responses adjacent to injury epicenter cavities following chronic human spinal cord injury. Exp Neurol. 2005;192(2):384–93.
53. Thuret S, Moon LD, Gage FH. Therapeutic interventions after spinal cord injury. Nat Rev Neurosci. 2006;7(8):628–43.
54. Yao L, Li Y, Knapp J, Smith P. Exploration of molecular pathways mediating electric field-directed Schwann cell migration by RNA-seq. J Cell Physiol. 2015;230(7):1515–24.
55. Li Y, Wang X, Yao L. Directional migration and transcriptional analysis of oligodendrocyte precursors subjected to stimulation of electrical signal. Am J Physiol Cell Physiol. 2015;309(8):C532–40.
56. Altman J, Das GD. Autoradiographic and histological evidence of postnatal hippocampal neurogenesis in rats. J Comp Neurol. 1965;124(3):319–35.
57. Altman J, Das GD. Postnatal neurogenesis in the Guinea-pig. Nature. 1967;214(5093):1098–101.
58. Caviness VS Jr. Time of neuron origin in the hippocampus and dentate gyrus of normal and reeler mutant mice: an autoradiographic analysis. J Comp Neurol. 1973;151(2):113–20.
59. Gueneau G, Privat A, Drouet J, Court L. Subgranular zone of the dentate gyrus of young rabbits as a secondary matrix. A high-resolution autoradiographic study. Dev Neurosci. 1982;5(4):345–58.
60. Stanfield BB, Trice JE. Evidence that granule cells generated in the dentate gyrus of adult rats extend axonal projections. Exp Brain Res. 1988;72(2):399–406.
61. Magavi SS, Leavitt BR, Macklis JD. Induction of neurogenesis in the neocortex of adult mice. Nature. 2000;405(6789):951–5.
62. Fallon J, Reid S, Kinyamu R, Opole I, Opole R, Baratta J, et al. In vivo induction of massive proliferation, directed migration, and differentiation of neural cells in the adult mammalian brain. Proc Natl Acad Sci U S A. 2000;97(26):14686–91.
63. Yoshimura N, Seki S, Novakovic SD, Tzoumaka E, Erickson VL, Erickson KA, et al. The involvement of the tetrodotoxin-resistant sodium channel Na(v)1.8 (PN3/SNS) in a rat model of visceral pain. J Neurosci. 2001;21(21):8690–6.
64. Yamamoto S, Nagao M, Sugimori M, Kosako H, Nakatomi H, Yamamoto N, et al. Transcription factor expression and notch-dependent regulation of neural progenitors in the adult rat spinal cord. J Neurosci. 2001;21(24):9814–23.
65. Madhavan L, Daley BF, Paumier KL, Collier TJ. Transplantation of subventricular zone neural precursors induces an endogenous precursor cell response in a rat model of Parkinson's disease. J Comp Neurol. 2009;515(1):102–15.
66. Elliott RC, Miles MF, Lowenstein DH. Overlapping microarray profiles of dentate gyrus gene expression during development- and epilepsy-associated neurogenesis and axon outgrowth. J Neurosci. 2003;23(6):2218–27.
67. Kojima T, Hirota Y, Ema M, Takahashi S, Miyoshi I, Okano H, et al. Subventricular zone-derived neural progenitor cells migrate along a blood vessel scaffold toward the post-stroke striatum. Stem Cells (Dayton, Ohio). 2010;28(3):545–54.
68. Nakatomi H, Kuriu T, Okabe S, Yamamoto S, Hatano O, Kawahara N, et al. Regeneration of hippocampal pyramidal neurons after ischemic brain injury by recruitment of endogenous neural progenitors. Cell. 2002;110(4):429–41.
69. McCaig CD, Song B, Rajnicek AM. Electrical dimensions in cell science. J Cell Sci. 2009;122.(Pt 23:4267–76.
70. Arocena M, Zhao M, Collinson JM, Song B. A time-lapse and quantitative modelling analysis of neural stem cell motion in the absence of directional cues and in electric fields. J Neurosci Res. 2010;88(15):3267–74.

References

71. Li Y, Wang PS, Lucas G, Li R, Yao L. ARP2/3 complex is required for directional migration of neural stem cell-derived oligodendrocyte precursors in electric fields. Stem Cell Res Ther. 2015;6:41.
72. Babona-Pilipos R, Droujinine IA, Popovic MR, Morshead CM. Adult subependymal neural precursors, but not differentiated cells, undergo rapid cathodal migration in the presence of direct current electric fields. PLoS One. 2011;6(8):e23808.
73. Nishino H, Hida H, Takei N, Kumazaki M, Nakajima K, Baba H. Mesencephalic neural stem (progenitor) cells develop to dopaminergic neurons more strongly in dopamine-depleted striatum than in intact striatum. Exp Neurol. 2000;164(1):209–14.
74. Shim JW, Park CH, Bae YC, Bae JY, Chung S, Chang MY, et al. Generation of functional dopamine neurons from neural precursor cells isolated from the subventricular zone and white matter of the adult rat brain using Nurr1 overexpression. Stem Cells (Dayton, Ohio). 2007;25(5):1252–62.
75. Zhang J, Calafiore M, Zeng Q, Zhang X, Huang Y, Li RA, et al. Electrically guiding migration of human induced pluripotent stem cells. Stem Cell Rev. 2011;7(4):987–96.
76. Feng JF, Liu J, Zhang XZ, Zhang L, Jiang JY, Nolta J, et al. Guided migration of neural stem cells derived from human embryonic stem cells by an electric field. Stem Cells (Dayton, Ohio). 2012;30(2):349–55.
77. Li Y, Weiss M, Yao L. Directed migration of embryonic stem cell-derived neural cells in an applied electric field. Stem Cell Rev. 2014;10(5):653–62.
78. Ozdemir M, Attar A, Kuzu I, Ayten M, Ozgencil E, Bozkurt M, et al. Stem cell therapy in spinal cord injury: in vivo and postmortem tracking of bone marrow mononuclear or mesenchymal stem cells. Stem Cell Rev. 2012;8(3):953–62.
79. Quertainmont R, Cantinieaux D, Botman O, Sid S, Schoenen J, Franzen R. Mesenchymal stem cell graft improves recovery after spinal cord injury in adult rats through neurotrophic and pro-angiogenic actions. PLoS One. 2012;7(6):e39500.
80. Zhao Z, Watt C, Karystinou A, Roelofs AJ, McCaig CD, Gibson IR, et al. Directed migration of human bone marrow mesenchymal stem cells in a physiological direct current electric field. Eur Cell Mater. 2011;22:344–58.
81. Kempermann G, Gast D, Kronenberg G, Yamaguchi M, Gage FH. Early determination and long-term persistence of adult-generated new neurons in the hippocampus of mice. Development. 2003;130(2):391–9.
82. Kamada M, Li RY, Hashimoto M, Kakuda M, Okada H, Koyanagi Y, et al. Intrinsic and spontaneous neurogenesis in the postnatal slice culture of rat hippocampus. Eur J Neurosci. 2004;20(10):2499–508.
83. Kaplan MS, Hinds JW. Neurogenesis in the adult rat: electron microscopic analysis of light radioautographs. Science. 1977;197(4308):1092–4.
84. Chenn A, McConnell SK. Cleavage orientation and the asymmetric inheritance of Notch1 immunoreactivity in mammalian neurogenesis. Cell. 1995;82(4):631–41.
85. Concha ML, Adams RJ. Oriented cell divisions and cellular morphogenesis in the zebrafish gastrula and neurula: a time-lapse analysis. Development. 1998;125(6):983–94.
86. Jaffe LF, Stern CD. Strong electrical currents leave the primitive streak of chick embryos. Science. 1979;206(4418):569–71.
87. Shi R, Borgens RB. Three-dimensional gradients of voltage during development of the nervous system as invisible coordinates for the establishment of embryonic pattern. Dev Dynam. 1995;202(2):101–14.
88. Hotary KB, Robinson KR. Endogenous electrical currents and voltage gradients in Xenopus embryos and the consequences of their disruption. Dev Biol. 1994;166(2):789–800.
89. Tuckett F, Morriss-Kay GM. The kinetic behaviour of the cranial neural epithelium during neurulation in the rat. J Embryol Exp Morphol. 1985;85:111–9.
90. Sausedo RA, Smith JL, Schoenwolf GC. Role of nonrandomly oriented cell division in shaping and bending of the neural plate. J Comp Neurol. 1997;381(4):473–88.

91. Barres BA, Raff MC. Proliferation of oligodendrocyte precursor cells depends on electrical activity in axons. Nature. 1993;361(6409):258–60.
92. Thompson HW, Malter JS, Steinemann TL, Beuerman RW. Flow cytometry measurements of the DNA content of corneal epithelial cells during wound healing. Invest Ophthalmol Vis Sci. 1991;32(2):433–6.
93. Song B, Zhao M, Forrester JV, McCaig CD. Electrical cues regulate the orientation and frequency of cell division and the rate of wound healing in vivo. Proc Natl Acad Sci U S A. 2002;99(21):13577–82.
94. Zhao M, Forrester JV. McCaig CD. A small, physiological electric field orients cell division. Proc Natl Acad Sci U S A. 1999;96(9):4942–6.
95. Pu J, Zhao M. Golgi polarization in a strong electric field. J Cell Sci. 2005;118(Pt 6):1117–28.
96. Fukata M, Nakagawa M, Kaibuchi K. Roles of rho-family GTPases in cell polarisation and directional migration. Curr Opin Cell Biol. 2003;15(5):590–7.
97. Zhao M, Pu J, Forrester JV, McCaig CD. Membrane lipids, EGF receptors, and intracellular signals colocalize and are polarized in epithelial cells moving directionally in a physiological electric field. FASEB J. 2002;16(8):857–9.
98. Zhao M, Agius-Fernandez A, Forrester JV, McCaig CD. Orientation and directed migration of cultured corneal epithelial cells in small electric fields are serum dependent. J Cell Sci. 1996;109(Pt 6):1405–14.
99. Ridley AJ, Schwartz MA, Burridge K, Firtel RA, Ginsberg MH, Borisy G, et al. Cell migration: integrating signals from front to back. Science. 2003;302(5651):1704–9.
100. Servant G, Weiner OD, Herzmark P, Balla T, Sedat JW, Bourne HR. Polarization of chemoattractant receptor signaling during neutrophil chemotaxis. Science. 2000;287(5455):1037–40.
101. Hall A. Rho GTPases and the actin cytoskeleton. Science. 1998;279(5350):509–14.
102. Fukata M, Kaibuchi K. Rho-family GTPases in cadherin-mediated cell-cell adhesion. Nat Rev Mol Cell Biol. 2001;2(12):887–97.
103. Nobes CD, Hall A. Rho GTPases control polarity, protrusion, and adhesion during cell movement. J Cell Biol. 1999;144(6):1235–44.
104. Etienne-Manneville S, Hall A. Integrin-mediated activation of Cdc42 controls cell polarity in migrating astrocytes through PKCzeta. Cell. 2001;106(4):489–98.
105. Etienne-Manneville S, Hall A. Cdc42 regulates GSK-3beta and adenomatous polyposis coli to control cell polarity. Nature. 2003;421(6924):753–6.
106. Kroschewski R, Hall A, Mellman I. Cdc42 controls secretory and endocytic transport to the basolateral plasma membrane of MDCK cells. Nat Cell Biol. 1999;1(1):8–13.
107. Luo L. Rho GTPases in neuronal morphogenesis. Nat Rev Neurosci. 2000;1(3):173–80.
108. Mellor H. Cell motility: Golgi signalling shapes up to ship out. Curr Biol. 2004;14(11):R434–5.
109. Preisinger C, Short B, De Corte V, Bruyneel E, Haas A, Kopajtich R, et al. YSK1 is activated by the Golgi matrix protein GM130 and plays a role in cell migration through its substrate 14-3-3zeta. J Cell Biol. 2004;164(7):1009–20.
110. Magdalena J, Millard TH, Etienne-Manneville S, Launay S, Warwick HK, Machesky LM. Involvement of the Arp2/3 complex and Scar2 in Golgi polarity in scratch wound models. Mol Biol Cell. 2003;14(2):670–84.
111. Magdalena J, Millard TH, Machesky LM. Microtubule involvement in NIH 3T3 Golgi and MTOC polarity establishment. J Cell Sci. 2003;116.(Pt 4:743–56.
112. Hayashi K, Kawai-Hirai R, Harada A, Takata K. Inhibitory neurons from fetal rat cerebral cortex exert delayed axon formation and active migration in vitro. J Cell Sci. 2003;116.(Pt 21:4419–28.
113. Jaffe LF. Electrophoresis along cell membranes. Nature. 1977;265(5595):600–2.
114. Poo M, Robinson KR. Electrophoresis of concanavalin a receptors along embryonic muscle cell membrane. Nature. 1977;265(5595):602–5.
115. Orida N, Poo MM. Electrophoretic movement and localisation of acetylcholine receptors in the embryonic muscle cell membrane. Nature. 1978;275(5675):31–5.
116. McLaughlin S, Poo MM. The role of electro-osmosis in the electric-field-induced movement of charged macromolecules on the surfaces of cells. Biophys J. 1981;34(1):85–93.

117. Brust-Mascher I, Webb WW. Calcium waves induced by large voltage pulses in fish keratocytes. Biophys J. 1998;75(4):1669–78.
118. Svitkina TM, Neyfakh AA Jr, Bershadsky AD. Actin cytoskeleton of spread fibroblasts appears to assemble at the cell edges. J Cell Sci. 1986;82:235–48.
119. Onuma EK, Hui SW. Electric field-directed cell shape changes, displacement, and cytoskeletal reorganization are calcium dependent. J Cell Biol. 1988;106(6):2067–75.
120. Henley J, Poo MM. Guiding neuronal growth cones using Ca2+ signals. Trends Cell Biol. 2004;14(6):320–30.
121. Mycielska ME, Djamgoz MB. Cellular mechanisms of direct-current electric field effects: galvanotaxis and metastatic disease. J Cell Sci. 2004;117(Pt 9):1631–9.
122. Jaffe LF, Nuccitelli R. Electrical controls of development. Annu Rev Biophys Bioeng. 1977;6:445–76.
123. Sato H, Ishii Y, Yamamoto S, Azuma E, Takahashi Y, Hamashima T, et al. PDGFR-beta plays a key role in the ectopic migration of neuroblasts in cerebral stroke. Stem Cells (Dayton, OH). 2016;34(3):685–98.
124. Nakada M, Nambu E, Furuyama N, Yoshida Y, Takino T, Hayashi Y, et al. Integrin alpha3 is overexpressed in glioma stem-like cells and promotes invasion. Br J Cancer. 2013;108(12):2516–24.
125. Tsai CH, Lin BJ, Chao PH. alpha2beta1 integrin and RhoA mediates electric field-induced ligament fibroblast migration directionality. J Orthop Res. 2013;31(2):322–7.
126. Pullar CE, Baier BS, Kariya Y, Russell AJ, Horst BA, Marinkovich MP, et al. beta4 integrin and epidermal growth factor coordinately regulate electric field-mediated directional migration via Rac1. Mol Biol Cell. 2006;17(11):4925–35.
127. Hannigan M, Zhan L, Li Z, Ai Y, Wu D, Huang CK. Neutrophils lacking phosphoinositide 3-kinase gamma show loss of directionality during N-formyl-met-Leu-Phe-induced chemotaxis. Proc Natl Acad Sci U S A. 2002;99(6):3603–8.
128. Polleux F, Whitford KL, Dijkhuizen PA, Vitalis T, Ghosh A. Control of cortical interneuron migration by neurotrophins and PI3-kinase signaling. Development. 2002;129(13):3147–60.
129. Zhao M, Song B, Pu J, Wada T, Reid B, Tai G, et al. Electrical signals control wound healing through phosphatidylinositol-3-OH kinase-gamma and PTEN. Nature. 2006;442(7101):457–60.
130. Spencer AG, Orita S, Malone CJ, Han MARHO. GTPase-mediated pathway is required during P cell migration in Caenorhabditis elegans. Proc Natl Acad Sci U S A. 2001;98(23):13132–7.
131. Liu JP. Jessell TM. A role for rhoB in the delamination of neural crest cells from the dorsal neural tube. Development. 1998;125(24):5055–67.
132. Alblas J, Ulfman L, Hordijk P, Koenderman L. Activation of Rhoa and ROCK are essential for detachment of migrating leukocytes. Mol Biol Cell. 2001;12(7):2137–45.
133. Causeret F, Hidalgo-Sanchez M, Fort P, Backer S, Popoff MR, Gauthier-Rouviere C, et al. Distinct roles of Rac1/Cdc42 and rho/Rock for axon outgrowth and nucleokinesis of precerebellar neurons toward netrin 1. Development. 2004;131(12):2841–52.
134. Hirose M, Ishizaki T, Watanabe N, Uehata M, Kranenburg O, Moolenaar WH, et al. Molecular dissection of the rho-associated protein kinase (p160ROCK)-regulated neurite remodeling in neuroblastoma N1E-115 cells. J Cell Biol. 1998;141(7):1625–36.
135. Yamazaki M, Miyazaki H, Watanabe H, Sasaki T, Maehama T, Frohman MA, et al. Phosphatidylinositol 4-phosphate 5-kinase is essential for ROCK-mediated neurite remodeling. J Biol Chem. 2002;277(19):17226–30.
136. Li L, El-Hayek YH, Liu B, Chen Y, Gomez E, Wu X, et al. Direct-current electrical field guides neuronal stem/progenitor cell migration. Stem Cells (Dayton, OH). 2008;26(8):2193–200.
137. Kim D, Kim S, Koh H, Yoon SO, Chung AS, Cho KS, et al. Akt/PKB promotes cancer cell invasion via increased motility and metalloproteinase production. FASEB J. 2001;15(11):1953–62.
138. Jiang H, Guo W, Liang X, Rao Y. Both the establishment and the maintenance of neuronal polarity require active mechanisms: critical roles of GSK-3beta and its upstream regulators. Cell. 2005;120(1):123–35.

139. Cantley LC. The phosphoinositide 3-kinase pathway. Science. 2002;296(5573):1655–7.
140. Meili R, Ellsworth C, Lee S, Reddy TB, Ma H, Firtel RA. Chemoattractant-mediated transient activation and membrane localization of Akt/PKB is required for efficient chemotaxis to cAMP in Dictyostelium. EMBO J. 1999;18(8):2092–105.
141. Jaffe LF, Nuccitelli R. An ultrasensitive vibrating probe for measuring steady extracellular currents. J Cell Biol. 1974;63(2 Pt 1):614–28.
142. Robinson KR. Electrical currents through full-grown and maturing Xenopus oocytes. Proc Natl Acad Sci U S A. 1979;76(2):837–41.
143. Robinson KR. Endogenous electrical current leaves the limb and prelimb region of the Xenopus embryo. Dev Biol. 1983;97(1):203–11.
144. Hotary KB, Robinson KR. Endogenous electrical currents and the resultant voltage gradients in the chick embryo. Dev Biol. 1990;140(1):149–60.
145. Hotary KB, Robinson KR. The neural tube of the Xenopus embryo maintains a potential difference across itself. Brain Res Dev Brain Res. 1991;59(1):65–73.
146. Hotary KB, Robinson KR. Evidence of a role for endogenous electrical fields in chick embryo development. Development. 1992;114(4):985–96.
147. Borgens RB, Shi R. Uncoupling histogenesis from morphogenesis in the vertebrate embryo by collapse of the transneural tube potential. Dev Dynam. 1995;203(4):456–67.
148. Metcalf MEM, Borgens RB. Weak applied voltages interfere with amphibian morphogenesis and pattern. J Exp Zool. 1994;268(4):323–38.
149. Chiang MC, Cragoe EJ Jr, Vanable JW Jr. Intrinsic electric fields promote epithelization of wounds in the newt, Notophthalmus viridescens. Dev Biol. 1991;146(2):377–85.
150. Sta Iglesia DD, Cragoe EJ Jr, Vanable JW Jr. Electric field strength and epithelization in the newt (Notophthalmus viridescens). J Exp Zool. 1996;274(1):56–62.
151. Zhao M, McCaig CD, Agius-Fernandez A, Forrester JV, Araki-Sasaki K. Human corneal epithelial cells reorient and migrate cathodally in a small applied electric field. Curr Eye Res. 1997;16(10):973–84.
152. Song B, Zhao M, Forrester J, McCaig C. Nerve regeneration and wound healing are stimulated and directed by an endogenous electrical field in vivo. J Cell Sci. 2004;117.(Pt 20:4681–90.
153. Reid B, Song B, McCaig CD, Zhao M. Wound healing in rat cornea: the role of electric currents. FASEB J. 2005;19(3):379–86.
154. Mendonca AC, Barbieri CH, Mazzer N. Directly applied low intensity direct electric current enhances peripheral nerve regeneration in rats. J Neurosci Methods. 2003;129(2):183–90.
155. Al-Majed AA, Brushart TM, Gordon T. Electrical stimulation accelerates and increases expression of BDNF and trkB mRNA in regenerating rat femoral motoneurons. Eur J Neurosci. 2000;12(12):4381–90.
156. Brushart TM, Hoffman PN, Royall RM, Murinson BB, Witzel C, Gordon T. Electrical stimulation promotes motoneuron regeneration without increasing its speed or conditioning the neuron. J Neurosci. 2002;22(15):6631–8.
157. Huang J, Lu L, Zhang J, Hu X, Zhang Y, Liang W, et al. Electrical stimulation to conductive scaffold promotes axonal regeneration and remyelination in a rat model of large nerve defect. PLoS One. 2012;7(6):e39526.
158. Borgens RB, Roederer E, Cohen MJ. Enhanced spinal cord regeneration in lamprey by applied electric fields. Science. 1981;213(4508):611–7.
159. Borgens RB, Blight AR, McGinnis ME. Behavioral recovery induced by applied electric fields after spinal cord hemisection in Guinea pig. Science. 1987;238(4825):366–9.
160. Fehlings MG, Tator CH, Linden RD. The effect of direct-current field on recovery from experimental spinal cord injury. J Neurosurg. 1988;68(5):781–92.
161. Borgens RB, Blight AR, Murphy DJ, Stewart L. Transected dorsal column axons within the Guinea pig spinal cord regenerate in the presence of an applied electric field. J Comp Neurol. 1986;250(2):168–80.
162. Shapiro S. A review of oscillating field stimulation to treat human spinal cord injury. World Neurosurg. 2014;81(5–6):830–5.

References

163. Wolf K, Wu YI, Liu Y, Geiger J, Tam E, Overall C, et al. Multi-step pericellular proteolysis controls the transition from individual to collective cancer cell invasion. Nat Cell Biol. 2007;9(8):893–904.
164. Beadle C, Assanah MC, Monzo P, Vallee R, Rosenfeld SS, Canoll P. The role of myosin II in glioma invasion of the brain. Mol Biol Cell. 2008;19(8):3357–68.
165. Wolf K, Muller R, Borgmann S, Brocker EB, Friedl P. Amoeboid shape change and contact guidance: T-lymphocyte crawling through fibrillar collagen is independent of matrix remodeling by MMPs and other proteases. Blood. 2003;102(9):3262–9.
166. Wagenaar-Miller RA, Engelholm LH, Gavard J, Yamada SS, Gutkind JS, Behrendt N, et al. Complementary roles of intracellular and pericellular collagen degradation pathways in vivo. Mol Cell Biol. 2007;27(18):6309–22.
167. Mohamed MM, cathepsins SBFC. Multifunctional enzymes in cancer. Nat Rev Cancer. 2006;6(10):764–75.
168. Cao L, Wei D, Reid B, Zhao S, Pu J, Pan T, et al. Endogenous electric currents might guide rostral migration of neuroblasts. EMBO Rep. 2013;14(2):184–90.
169. Hubbeling D. Registering findings from deep brain stimulation. JAMA. 2010;303(21):2139–40. author reply 40
170. Follett KA, Weaver FM, Stern M, Hur K, Harris CL, Luo P, et al. Pallidal versus subthalamic deep-brain stimulation for Parkinson's disease. N Engl J Med. 2010;362(22):2077–91.
171. Rothlind JC, York MK, Carlson K, Luo P, Marks WJ Jr, Weaver FM, et al. Neuropsychological changes following deep brain stimulation surgery for Parkinson's disease: comparisons of treatment at pallidal and subthalamic targets versus best medical therapy. J Neurol Neurosurg Psychiatry. 2015;86(6):622–9.
172. Robinson KR, Messerli MA. Electric embryos: the embryonic epithelium as a generator of developmental information. In: McCaig CD, editor. Nerve growth and nerve guidance. Portland Press; 1996. p. 131–50.
173. Ingvar S. Reactions of cells to the galvanic current in tissue culture. Proc Soc Exp Biol Med. 1920;17:198–9.

Chapter 6
Vascularization in the Spinal Cord: The Pathological Process in Spinal Cord Injury and Therapeutic Approach

Hien Tran and Li Yao

Abstract Spinal cord injury (SCI) often results in disruption of the vascular structures at the area of damage in the case of primary injury. The secondary injury of SCI is the result of pathophysiological changes such as excitotoxicity, inflammation, ischemia, and oxidative stress. The central nervous system (CNS) retains limited ability to angiogenic potential in order to restore the vascular network following SCI. Researches showed the formation of new blood vessels from mesenchymal cells. Multiple approaches such as the use of bio-scaffolds, cell transplantation, molecular therapy, electrical stimulation, and hypothermia have been used to explore the potential of assisting the endogenous revascularization. This chapter attempts to summarize the advance of the research on regeneration of spinal cord vascular networks and functions.

Keywords Spinal cord · Vascularization · Extracorporeal shock wave therapy · Hypothermia · Biomaterial scaffolds · Reconstruction of vascular structure · Molecular therapy · Ischemia · Growth factors

6.1 Vascular Structure of Spinal Cord

Spinal cord injury (SCI) is the structural and functional disruption of the spinal cord due to multiple causes—both traumatic and atraumatic. More than three million people are living with SCI, and more than 12,000 new cases of SCI are encountered each year in the United States. SCI greatly and negatively affects daily routines, including difficulty in ambulation, sexual arousal, and more. During SCI, the vascular network of the spinal cord is heavily affected, initiating primary injury, which then turns into a series of additional damages known collectively as secondary injury. For a short amount of time up until the 1960s, it was generally accepted that the central nervous system (CNS) had absolutely no potential for regeneration [1]. However, it was soon discovered that regions of the brain do maintain the ability of neuroregeneration. Likewise, the CNS also retains the ability to regenerate the vascular network although with minimal success in the majority of cases due to multiple mechanisms that prevent recovery of function.

The study of SCI, its effects on the vascular and neural networks of the spinal cord, and attempts to remedy the condition are ongoing, with promises recently arising from the use of bio-scaffolds, cell transplantation, molecular therapy, electrical stimulation, and hypothermia, along with classical approaches such as the use of corticosteroids, which help reduce inflammation and trigger regeneration. This chapter attempts to utilize the published data on vascularization following SCI to summarize the advance of the research on regeneration of spinal cord vascular networks and functions.

In adults the thoraco-abdominal aorta, the vertebra-subclavian arteries, and occasionally the internal iliac arteries comprise the main sources of the arterial supply of the spinal cord [3]. The vascularization of the spinal cord is divided into three major spinal arterial territories including cervicothoracic area, midthoracic area, and thoracolumbar area [4, 5].

Both extrinsic and intrinsic arteries provide blood supply to the spinal cord. The extrinsic arteries are comprised of the anterior spinal artery, the posterior spinal artery, and the pial network, with the radiculomedullary artery supplying the anterior spinal artery and the radiculopial artery supplying the posterior spinal artery and the pial network [6]. The intrinsic arteries are divided into the central and peripheral arteries. The central arteries branch from the anterior spinal artery and provide centrifugal blood flow to most of the gray matter and part of the white matter [4, 5]. The central arteries vary in size and number. Arteries in the cervical and lumbar regions exist in greater numbers and cross-section diameters, and thus provide a larger blood supply compared to those in the thoracic region [7]. By contrast, the peripheral arterial supply to the cervical and thoracic regions is rich and to the lumbar region is poor [5]. The peripheral system branches from the posterior spinal artery and pial arterial plexus and provides the centripetal blood supply to the dorsal part of the spinal cord and peripheral white matter [5, 8]. The central and peripheral systems fuse in a complex network of terminal capillary beds. A capillary bed that is about five times as dense supplies the gray matter, compared to that which supplies the white matter [5]. This is most likely the result of gray matter, which is the location of neuronal cell bodies, requiring more energy and oxygen than white matter, which is composed of axon tracts [7].

Even though the central and peripheral arteries appear to supply the spinal cord in a complementary fashion, they do not necessarily maintain a constant blood supply throughout the spinal cord because their blood flows are in different directions. Therefore, certain regions are more prominently vascularized than others. Unsurprisingly, it has been noted that possessing a greater degree of vascularization allows certain regions to avoid infarction, which is the death of tissues due to ischemia. When a region possesses a dense network of blood vessels, dysfunction of one blood vessel can be alleviated by the numerous nearby blood vessels that can aid in continuing the blood supply to tissues.

The venous system in the spinal cord is slightly less complex than that in the arterial system. The intrinsic veins perforate symmetrically from the axis of the spinal cord, distribute uniformly throughout the entire cord, and connect to a pial network superficially on the cord [6, 9]. The anterior sulcal vein drains into the anterior spinal vein, and the posterior sulcal vein drains into the posterior spinal vein. The extrinsic veins include the anterior spinal vein, posterior spinal vein, and pial venous plexus [6]. The veins of the spinal cord drain blood in the opposite direction of the arterial workings.

6.2 Damage of Vasculature in Spinal Cord Injury

Traumatic SCI can arise from a variety of situations including motor vehicle accidents, gunshot wounds, falls, and surgical complications, among others [10]. These can cause compression, excessive extension, excessive flexion, and other deformations of the spinal cord. It was reported that the pia mater and the axons of fully developed spinal cords have limited plasticity; therefore, the cord can flatten and decrease in diameter in the case of compression and changes in the cord's length [3]. Atraumatic causes of SCI include inflammatory and autoimmune processes [11]. There is no universal timeline for the events occurring after SCI because the manifestations are multifactorial. A SCI rarely results in complete severance of the spinal cord, but instead often results in disruption of the vascular and cellular structures at the area of damage in the case of primary injury. The secondary injury is the result of pathophysiological changes such as excitotoxicity, inflammation, ischemia, and oxidative stress.

The primary injury involves the disruption of small vascular networks, which often result in a minor hemorrhage at the SCI epicenter [12]. Studies have shown that the vascular disruption initially occurs near the central canal in the gray matter and spreads to white matter structures after several hours [8]. During the first 2 days, staining by adhesion molecules such as PECAM provides evidence of significant endothelial cell loss proximal to the injury epicenter [13]. Damage to the vasculature at the site of injury results in decreased blood supply to the neural cells such as neurons and astrocytes. Since neurons have a high need for blood for its high level of metabolism, even transient periods of ischemia and hypoxia may be disastrous and lead to their death. However, primary injury usually does not cause a significant amount of cell death in comparison to total cell death [14].

Secondary injury occurs mostly 24 h post-SCI and involves a variety of mechanisms including excitotoxicity, inflammation, and ischemia. SCI also induces oxidative stress due to mitochondrial dysfunction after release of excitatory amino acids and dysfunction in calcium homeostasis [15]. Phagocytic cells that are released following SCI also play a role in the creation of reactive oxygen species (ROS) because they consume an increased amount of oxygen leading to the increased generation of superoxide [15]. Overall, vascular density progressively decreases, and the site of injury retains almost no detectable vessels after 48 h [13, 14].

At the site of a spinal cord injury, there is often one region that is incapable of revascularization and suffers from complete necrosis with hemorrhage and neuronal death. Surrounding this region is another region that can revascularize within 7 days [16]. Sizes of these regions vary among different types of SCI [16]. Generally, it is suggested that clinical interventions should be initiated within 3 days to 1 week and should target early revascularization [17]. One study examined the dynamic angiogenic response initiated after a moderately severe clip-compression injury in rats [17]. The study quantified proliferating endothelial cells at the epicenter and multiple distances rostral and caudal to the epicenter through Ki67 antibody detection. The researchers observed that there was angiogenesis starting on day 3, reaching maximal potential on day 5, and ending on day 10 post-injury.

Human studies of events occurring post-SCI are rare due to the difficulty in obtaining cord specimen from humans. One study involving human SCI from C3 to T11 inclusive attempted to measure the amounts of the angiogenic proteins—Ang-1, Ang-2, and angiogenin—in the cerebrospinal fluid and serum samples in both SCI patients and a control group. The researchers collected only one cerebrospinal fluid (CSF) and serum sample from the control group and assumed that the concentrations of these proteins would stay unchanged over time. The multiple CSF samples collected from SCI patients showed that the Ang-1 level increased within the first 12 h after injury and then returned to baseline [8]. However, the Ang-2 level was shown to be elevated approximately 36 h after injury and remained elevated for the rest of the 120-h study. In regard to angiogenin, its concentration in the CSF started to decrease around 36 h, with the greatest difference compared to the control group at 72 h. Angiogenin in the serum appeared to start increasing between 60 and 80 h post-injury although the difference was minimal compared to the control group. No neurologic manifestations after 6 months or 1 year were correlated with the levels of Ang-1 and Ang-2 [8]. Matrix metalloproteinases (MMP) are enzymes that can break down the extracellular matrix and are important for cell migration. Their role in SCI has varied through different studies, depending on which type of MMP is being examined. MMP-2 has been observed to be increasing between 7 and 14 days post-injury. Deficiency in MMP-2 in rat models resulted in increased white matter damage and increased impairment in open-field locomotion, rotarod performance, and grid walking, suggesting that MMP-2 might be beneficial after SCI [18]. The increase of another type of MMP, MMP-9, during the acute phase of injury leads to increased blood-spinal cord barrier (BSCB) permeability, decreased attenuation of macrophage infiltration, and decreased locomotor functional recovery, suggesting that MMP-9 exacerbates SCI [19].

6.3 Natural Process of Revascularization and Remedies Post–SCI

The human body possesses mechanisms for self-repair and recovery after spinal cord insults although its efficacy has been less than desirable. Consequently, a significant population of SCI patients ends up with long-term disabilities or even death. Regardless, observation of the body's reactions after SCI is important for the emergence of exogenous intervention.

Inflammation in the spinal cord is different than inflammation in the brain, despite both organs being part of the central nervous system. Generally, there is a greater degree of acute inflammation in the spinal cord compared to the cerebral cortex after similar injury in a rat [20]. It was observed that after 1 day of injury, the lesion site in the spinal cord contained twice as many neutrophils as in the brain, and the amount of neutrophil infiltration to nearby parenchyma in the spinal cord was quadruple that in the brain. One similarity between the two CNS compartments lies in the decrease of neutrophils in both after several days. Researchers also noticed that the microglia/macrophage and lymphocyte infiltration and activation were more widespread in the spinal cord [20]. Non-traumatic microinjection of the cytokines IL-1β or TNFα into rat spinal cord showed greater recruitment of neutrophils and CNS macrophages in the spinal cord than in the brain. It also showed increased lymphocyte infiltration into the spinal cord but not the brain [21]. Injection of TNFα showed neutrophils and macrophages infiltrating into the spinal cord, whereas only macrophages infiltrate into the brain [21].

The blood-spinal cord barrier is a layer composed of endothelial cells with tight junctions between them and a complex negatively charged glycoprotein-rich glycocalyx on the luminal front to control the entrance and exit of various materials into and out of the central nervous system [18]. The integrity of the BSCB is extremely important for normal neurological function. SCI results in the disruption of the BSCB, causing a temporary loss of the negative charge of the glycocalyx, which is responsible for repulsing plasma proteins—such as cells that yield inflammation—that possess a similar charge [22]. This allows for the uncontrolled passage of cytotoxic molecules such as calcium, excitatory amino acids, free radicals, erythrocytes, and inflammatory mediators into the site of injury, all of which may contribute to secondary injury. BSCB dysfunction has mostly been reported as a temporary condition; however, chronic BSCB disruption following SCI has been observed several months after the initial injury [21].

The body also appears to retain angiogenic potential following SCI. Around 3–4 days after injury and up to about 1 week, angiogenesis, the process of new blood vessels sprouting outward from existing blood vessels, was observed [23]. Revascularization, the formation of new blood vessels from mesenchymal cells, has been observed to increase baseline levels and in some cases even more than 500% at the one-week mark. These newly formed vessels, however, are not associated with neurons, astrocytes, or pericytes [14]. The roles of astrocytes and pericytes have been emphasized to be extremely important on neural function within the nervous system. Therefore, despite the vehicle for blood supply being replenished, it would not be meaningful if astrocytes and pericytes cannot attach to and interact with them. Another problem involves the rapid degradation of these new blood vessels at the injury epicenter after about 2 weeks, while the number of astrocytes in this area keeps increasing [23]. This suggests that there is a need for maintenance and protection of the newly formed blood vessels and a need to facilitate interaction between the vessels and cells of the CNS in addition to kickstarting the restoration of those blood vessels. Interestingly, BSCB permeability increases during this period of angiogenesis. It could be explained that there needs to be a transient disruption in vascular structures while vascular remodeling, such as in the case of angiogenesis, is in process.

6.4 Therapeutic Approaches to Reconstruction of Vascular Structure Following SCI

It is generally accepted that regrowth of the vascular network status post-SCI has great potential for functional recovery [28]. It is worth noting that in order to achieve functional recovery after SCI, it is necessary to restore those components that have been damaged, which include both vascular and neural networks. Studies that have attempted to restore only one or the other may or may not find a strongly positive correlation between restoration of either network and functional recovery. The recovery of blood vessels must occur along with the recovery of neural cells, such as with axonal growth and restoration of functional connection. It is a common practice for researchers to look only at revascularization and assume that the functional recovery is due mainly to the regrowth of vascular structures; therefore, some studies do not assess the status of the neural network (Table 6.1).

6.4.1 Therapeutic Molecules

Therapeutic molecules, either through intravenous injections or delivery using a cellular vehicle, can help relieve inflammation and promote restoration of the vascular and neural networks. Angiopoetin-1 has been one of the most experimental molecules, either alone or in combination with other molecules, to relieve and reverse SCI. It is essential to protect the vasculature at, and in proximity to, the site of injury to minimize damage and allow for better recovery. Protection of the BSCB and white matter has correlated with the number of perfused blood vessels and is a promising target. Experimentation on female C57BL/6 mice using a combination of C16, which is an integrin-binding peptide, and Ang-1 was shown to reduce blood vessel loss at the injury site as early as 24 h [24]. White matter loss was also limited 7 and 42 days post-injury [24] although there was only a small difference in terms of improvement at 42 days compared to 7 days. Inflammation within the first 24 h was also limited, suggesting that if these molecules can be introduced during the acute phase of the SCI, there is the potential to effectively treat SCI. More interestingly, the treatment in this experiment was done 4 h after the initial injury, which suggests that there is a realistic chance that clinical application could be possible [24]. The role and usage of Ang-2 in SCI are more complicated and controversial. This protein acts as an antagonist to Ang-1, competing for the same receptor, the endothelial tyrosine kinase receptor (Tie-2 receptor). Ang-1 is down-regulated during SCI, which is correlated with decreased integrity of the vascular network. Ang-2 has been shown to be up-regulated following SCI and has been thought of as a possible contributor to vascular breakdown [8]. However, the up-regulation of Ang-2 has not been associated with the breakdown of the vasculature [2]. In fact, an increase in Ang-2 has been shown to correlate with improved vasculature regeneration in male Sprague-Dawley rats.

6.4 Therapeutic Approaches to Reconstruction of Vascular Structure Following SCI   117

Table 6.1 Reconstruction of vascular structure following SCI

Treatment	Materials	Animals	Approach	Outcome	Refs.
Molecular treatment	VEGF, angiostatin	Sprague-Dawley rats	Injection of VEGF165, or angiostatin into the site of contusion SCI	VEGF increases amount of rescued tissues and blood vessels near injury site and decreased apoptosis. Angiostatin injection did not produce any statistically significant difference compared to control	[29]
Molecular treatment	VEGF	Fischer 344 rats	Injection of rhVEGF165 into injured spinal cord 72 h post-injury	Increased microvascular permeability and increased leukocyte infiltration into the spinal cord	[30]
Molecule/biomaterial	VEFG + FGF-2 loaded on PLG microspheres	Long-Evans rats	VEGF and FGF-2 were encapsulated by PLG microspheres and injected into T9-T10 hemisection SCI	Positive infiltration of endothelial cells into the bridge at 6 weeks after the injury and histological study showed circular-shaped blood vessels adjacent to the bridge in higher numbers compared to control	[35]
Molecular treatment	VEGF and Angiopoetin-1	Sprague-Dawley rats	Adeno-associated virus (AAV) delivery of Angiopoeitin-1 and VEGF165 into T7 SCI	Decreased hyperintense lesion volume, suggesting decreased edema and demyelination. No difference on hypointense lesion volume, indicator of necrosis and hemorrhage. Mild restoration of BSCB vascular stability. Inflammation was not exacerbated and was similar to control group	[26]

(continued)

Table 6.1 (continued)

Treatment	Materials	Animals	Approach	Outcome	Refs.
Molecular treatment	Angiopoeitin-1 and C16	C57BL/6 mice	Daily IV injections of Ang-1 mimetic and C16 for contusion SCI in rats	Blood vessels and white matter at the injury penumbra were rescued. Inflammation was decreased	[24]
Molecular treatment	Prednisolone	Adult mongrel cats	Methylprednisolone sodium succinate (MPSS) was injected through the IV into cats 30 min after compression injury of the spinal cord in the lumbar region. MPSS was also given through the IV 20 min prior to the injury	Inflammation and tissue necrosis was reduced, evidenced by the reduced rate of lipid hydrolysis and tissue level of free fatty acids and arachidonate, signifying that cell membrane hydrolysis was reduced	[31]
Biomaterial graft	PLGA 504 oriented outer scaffold + microporous PEG/PLL hydrogel inner scaffold	Female Sprague-Dawley rats	Scaffold combo carrying NPCs and ECs grafted into site of the rat's thoracic spinal cord of hemisection	The researchers observed angiogenesis that involves a three- to five-fold increase in the amount of blood vessels that were functional at the SCI penumbra after 8 weeks	[33]
Biomaterial graft	Alginate-fibrinogen combination	Female long-Evans rats	VEGF nanoparticle-loaded alginate-fibrinogen hydrogel was injected into the site of the rat's thoracic spinal cord of hemisection	Angiogenesis was observed and spinal cord plasticity was increased. However, functional correlation was absent	[27]
Biomaterial graft	Alginate-g-pyrrole + VEGF-encapsulated PLGA microspheres	Fertilized chicken eggs	Implantation of HRP/H2O2-activated alginate-g-pyrrole hydrogel encapsulated with PLGA microspheres containing VEGF onto chicken chorioallantoic membranes	The release of VEGF was better sustained, and local neovascularization was more potent compared with Ca2+ cross-linked alginate and alginate-g-pyrrole hydrogel in terms of number and size of blood vessels	[34]

(continued)

6.4 Therapeutic Approaches to Reconstruction of Vascular Structure Following SCI   119

Table 6.1 (continued)

Treatment	Materials	Animals	Approach	Outcome	Refs.
Extracorporeal shockwave therapy	N/A	Adult female Sprague-Dawley rats	ESW was administered to 2 spots on the injured cord for 3 weeks at the frequency of 3 times/week	Significant locomotor improvement compared with the SCI group at 7, 35, and 42 days	[41]
Systemic hypothermia	N/A	Adult rats	Systemic hypothermia at 33 °C 30 min following compression SCI for 4 h	Increased VEGF expression and increased endothelial cell number	[45]
Systemic hypothermia	N/A	Adult rats	Systemic hypothermia at 33 °C 5 min after cervical contusion SCI for 4 h	Improvement in limb function and BBB scores between 1 and 3 weeks. Higher amount of rescued motor neurons and preserved axons at the site of injury in histological studies	[46]

Researchers tend to believe that a higher dose of vascular endothelial growth factor (VEGF) would potentially allow for increased stimulation of angiogenesis; therefore, effort is always made to more efficiently deliver VEGF to the site of injury. One study targeted and up-regulated the VEGF gene using hypoxia in order to achieve increased VEGF production. Utilizing the Epo enhancer and the RTP801 promoter, after transplantation into the rat's spinal cord, the system was shown to positively influence the expression of VEGF and induce neuronal growth and revascularization [25]. Injected exogenous VEGF can be useful in alleviating the damage caused by secondary reactions although it must be injected into the spinal cord at the right time. A study involving male Sprague-Dawley rats have shown that when VEGF is injected into the site of SCI immediately after the injury, there is microvascular regeneration, improvement in functional outcome, and a decrease in secondary damage [29]. However, when VEGF is injected 72 h after SCI in Fischer 344 rats, the microvascular networks become more permeated allowing leukocytes to enter the spinal cord and overall worsen the injury [30]. Following SCI, levels of VEGF and Ang-1 decrease drastically. A study attempting to supplement these two factors using viral vectors based on the adeno-associated virus (AAV) on adult male Sprague-Dawley rats showed decreased lesion volume, improved Basso, Beattie, and Bresnahan (BBB) scores 56 days after injury and minimal exacerbation of the infiltration of microglia into the site of injury [26].

Another therapy that has been controversial but has been used nevertheless is anti-inflammatory drugs, particularly methylprednisolone. One study in which methylprednisolone sodium succinate (MPSS) was injected intravenously into cats

20 min prior to compression injury of the spinal cord in the lumbar region and also administered intravenously 30 min after injury showed promising results [31]. Inflammation and tissue necrosis were reduced, evidenced by the reduced rate of lipid hydrolysis and tissue level of free fatty acids and arachidonate, signifying that cell membrane hydrolysis was reduced [31].

6.4.2 Biomaterial Scaffolds

Biomaterial scaffolds can assist in treating SCI either by acting as a vehicle for transportation of essential compounds or by acting as a scaffold on which new cells can develop. Most biomaterial scaffolds are designed to match the cellular environment characteristics that can decrease the risk of chronic inflammation and other complications. They are also degradable in the body, eliminating the need for surgery to remove the scaffolds once treatment ends. The use of biomaterial scaffolds that target vascular regeneration in the spinal cord has not been addressed as frequently by researchers as scaffolds used for neuronal regeneration.

Porous scaffolds made of biodegradable and biocompatible block copolymer of poly-lactic-co-glycolic acid (PLG, PLGA) and poly-L-lysine (PLL) scaffolds can be safely implanted into an injured primate spinal cord, and these implants can allow for tissue remodeling, gray matter protection, and improvement recovery of locomotion in Old World African green monkeys that have undergone a complete lateral thoracic hemisection spinal cord injury [32]. In another study, the implantation of a PLGA scaffold containing a co-culture of neural progenitor cells (NPCs) and endothelial cells attempted to create a microenvironment that allows for vascular structure growth. The researchers observed angiogenesis that showed a three- to five-fold increase in the amount of blood vessels that were functional at the lesion epicenter after 8 weeks [33]. They also found that approximately 50% of the vessels were positive for the endothelial barrier antigen, indicating that these vessels could have been formed in the fashion of the blood-spinal cord barrier [33]. Results suggest that by utilizing a scaffold, there is potential to create functional vascular structures that can serve as both blood vessels and a blood-spinal cord barrier. Studies using PLGA scaffolds have been used not only for angiogenesis but also for neovascularization, the process of de novo blood vessel formation compared to blood vessels sprouting from existing ones. Neovascularization was studied using fertilized chicken eggs using alginate-g-pyrrole hydrogel containing VEGF-encapsulated PLGA microspheres. The hydrogel was activated by HRP/H2O2. More potent neovascularization was observed, both in terms of number and size of blood vessels [34]. Another study that utilized VEGF and fibroblast growth factor-2 (FGF-2) that were encapsulated by PLG microspheres and injected locally into the SCI site to act as a bridge found positive infiltration of endothelial cells into the bridge at 6 weeks after injury, and a histological study showed circular-shaped blood vessels adjacent to the bridge in higher numbers compared to the control [35].

Alginate-fibrinogen hydrogel with free VEGF and VEGF nanoparticles was injected into female Long-Evans rats at the site of the thoracic spinal cord hemisection. The researchers observed increased angiogenesis with endothelial cell infiltration into the lesion epicenter. They also observed increased spinal cord plasticity with great neurite growth around the lesion, evidenced by greater beta-III tubulin and GAP43 staining. However, functional recovery was not observed [27]. In another study, neovascularization and neurite growth were observed in adult female Sprague-Dawley rat models through the transplantation of porous collagen scaffolds containing neurotrophic factors (BDNF and NT-3), compounds that induce neutralization of myelin inhibitory molecules (Nogo, ephrinB3, and sema4D), and injection of cAMP into the gray matter near the lesion site to activate neuronal intrinsic mechanisms proved to be the most effective approach in reducing cavitation volumes and increasing revascularization, axonal regeneration, neuronal generation, and overall locomotion recovery [36].

6.4.3 Extracorporeal Shock Wave Therapy

Other novel therapeutic techniques that have not been explored closely but have shown promising results include extracorporeal shock wave therapy (ESWT) and hypothermia. ESWT, often synonymous with functional extracorporeal shock wave therapy (fESWT), is a nonsurgical therapeutic technique based on electrohydraulic, electromagnetic, or piezoelectric sources, that has been used for various conditions [37]. For example, extracorporeal shock wave lithotripsy has been used and shown to be efficient in treating renal calculi [38], tennis elbow for its analgesic properties [39], and various bone and tendon problems [40]. fESWT has also been used in myocardial infarction patients and seems to induce angiogenesis locally at the coronary microvasculature. The use of ESWT in adult female Sprague-Dawley rat models has shown to increase angiogenesis and recovery of functions in mice afflicted with myocardial ischemia, myocardial infarction, and peripheral artery disease. It also allows for vascular structure formation at the tendon-bone junction. A study by Yamaya et al. [41] utilizing this technology on spinal cord injury applied ESWT emitted by a DUOLITH-SD1 shock wave system to two spots on an injured spinal cord three times per week for 3 weeks following injury and after wound closure. The shock wave was shown to not damage normal tissues because both control groups, one with exposure to ESWT and one without, showed similar healthy behavior. The experimental group demonstrated significant locomotor improvement compared with the control group at 7, 35, and 42 days. Another study was performed on patients with chronic neurogenic heterotrophic ossification (NHO) in the knee and hip following traumatic brain injury (TBI) [42]. All participants received EWST four times over the course of 8 weeks to the affected knee or hip. The results, based on visual inspection, showed that ESWT seemed to significantly improve flexion and extension of the hip and improved flexion of the knee. There was no significant difference in knee extension performance over time. This study was significant in that it made use of human subjects. However, due to this fact, the number of participants was small and not equally distributed between males and females—more males than females—although this represents the general trend in population that usually present with such problems. Another limitation was the lack of a control group, which the author noted was difficult with such a population of patients due to their tendency to present with profound cognitive impairment [42]. In a separate study, the same researchers found that ESWT also reduced pain, indicated by the reduction in faces rating scale (FRS) pain scores [43]. This study had the same limitations as above, with a small population and lack of a control group. Results were compared within the respective case against the baseline characteristics of the conditions that were assessed at the beginning of the study [43]. A controversial and less-studied form of ESWT, called radial ESWT (rEWST), utilizes waves that are at a lower intensity and speed compared to fESWT, and are sometimes referred to as pressure waves instead of shock waves. In a study of middle cerebral artery occlusion (MCAO) and 12 days of rESWT, researchers noted an increase in cerebral blood flow, increase in VEGF, and decrease in brain infarct volume [44]. They also observed that levels of neuron-specific enolase and nestin increased, which suggests that there was an increase in the number of neurons and neural stem cells. Up-regulation of the Wnt and beta-catenin expression suggest that neural stem cells proliferated and differentiated through the Wnt/beta-catenin pathway through rESWT [44].

6.4.4 Hypothermia

Hypothermia is utilized at different levels ranging from sub-30 °C to 34 °C. However, several studies have shown that profound hypothermia (<30 °C) usually results in more damage in the brain compared to moderate hypothermia (30–32 °C) and modest hypothermia (32–34 °C) [45]. Therefore, it is generally accepted that modest to moderate hypothermia possess the greatest potential for minimal damage [45]. In a rat compression SCI model, hypothermia at 33 °C administered systemically after 30 min post-injury for 4 h led to an increase in VEGF expression and an increase in the number of endothelial cells [46]. Another study with a similar setup [47] initiated hypothermia (33 °C) systemically in the rat within 5 min after a cervical spinal cord contusion injury for 4 h and then rewarmed the animal at 1 °C per hour. The results showed improvement in limb function and BBB scores between 1 and 3 weeks. More importantly, the histological assessment of a segment of the injured spinal cord showed higher amounts of rescued ventral motor neurons and preserved axons. There was also greater rescue of the gray and white matter. This suggests that hypothermia can potentially be used to preserve vascular and neural networks and promote angiogenesis. Hypothermia can work in combination with other therapies as well [48]. Wang et al. [49] performed bone marrow mesenchymal stem cell transplantation along with hypothermia at 33–35 °C in rat models and observed no cavity formation at the site of SCI, but instead a significantly greater number of preserved neurons. Results suggest that hypothermia can work alone and in combination with other therapies and can both promote decreased damage as well as increased preserved and regrowth of important components.

In conclusion, spinal cord injuries can arise from both traumatic and atraumatic origins and can greatly affect the delicate vascular network in the spinal cord. After injury, disruption of the vascular network is followed by ischemia, BSCB disruption, excitotoxicity, inflammation, and ROS, which further exacerbate the damage. Revascularization can be seen for several days but appears to ultimately be inefficient due to the inability of the new vascular structures to interact with cells of the neural network and the failure of structures to remain intact. Multiple approaches have been used to explore the potential of assisting with endogenous revascularization through the use of biomaterials, cell transplantation, essential therapeutic molecules, extracorporeal shock wave, and hypothermia.

References

1. Tsintou M, Dalamagkas K, Seifalian AM. Advances in regenerative therapies for spinal cord injury: a biomaterials approach. Neural Regen Res. 2015;10(5):726–42. https://doi.org/10.4103/1673-5374.156966.
2. Durham-Lee JC, Wu Y, Mokkapati VU, Paulucci-Holthauzen AA, Nesic O. Induction of angiopoietin-2 after spinal cord injury. Neuroscience. 2012;202:454–64. https://doi.org/10.1016/j.neuroscience.2011.09.058.
3. Tveten L. Spinal cord vascularity. I. Extraspinal sources of spinal cord arteries in man. Acta Radiol Diagn (Stockh). 1976;17(1):1–16.
4. Lazorthes G, Gouaze A, Zadeh JO, Santini JJ, Lazorthes Y, Burdin P. Arterial vascularization of the spinal cord. Recent studies of the anastomotic substitution pathways. J Neurosurg. 1971;35(3):253–62.
5. Martirosyan NL, Feuerstein JS, Theodore N, Cavalcanti DD, Spetzler RF, Preul MC. Blood supply and vascular reactivity of the spinal cord under normal and pathological conditions. J Neurosurg Spine. 2011;15(3):238–51. https://doi.org/10.3171/2011.4.Spine10543.
6. Miyasaka K, Asano T, Ushikoshi S, Hida K, Koyanagi I. Vascular anatomy of the spinal cord and classification of spinal arteriovenous malformations. Interv Neuroradiol. 2000;6(Suppl 1):195–8. https://doi.org/10.1177/15910199000060s131.
7. Turnbull IM. Microvasculature of the human spinal cord. J Neurosurg. 1971;35(2):141–7. https://doi.org/10.3171/jns.1971.35.2.0141.
8. Ng MT, Stammers AT, Kwon BK. Vascular disruption and the role of angiogenic proteins after spinal cord injury. Transl Stroke Res. 2011;2(4):474–91. https://doi.org/10.1007/s12975-011-0109-x.
9. Amato ACM, Stolf NAG. Anatomy of spinal blood supply. J Vasc Brasileiro. 2015;14(3):248–52.
10. Chen Y, Tang Y, Vogel LC, DeVivo MJ. Causes of spinal cord injury. Topics in Spinal Cord Injury Rehabilitation. 2013;19(1):1–8.
11. Grassner L, Marschallinger J, Dünser MW, Novak HF, Zerbs A, Aigner L, Trinka E, Sellner J. Nontraumatic spinal cord injury at the neurological intensive care unit: spectrum, causes of admission and predictors of mortality. Ther Adv Neurol Disord. 2016;9(2):85–94. https://doi.org/10.1177/1756285615621687.

12. Dray C, Rougon G, Debarbieux F. Quantitative analysis by in vivo imaging of the dynamics of vascular and axonal networks in injured mouse spinal cord. Proc Natl Acad Sci U S A. 2009;106(23):9459–64. https://doi.org/10.1073/pnas.0900222106.
13. Whetstone WD, Hsu J-YC, Eisenberg M, Werb Z, Noble-Haeusslein LJ. Blood-spinal cord barrier after spinal cord injury: relation to revascularization and wound healing. J Neurosci Res. 2003;74(2):227–39. https://doi.org/10.1002/jnr.10759.
14. Casella GT, Marcillo A, Bunge MB, Wood PM. New vascular tissue rapidly replaces neural parenchyma and vessels destroyed by a contusion injury to the rat spinal cord. Exp Neurol. 2002;173(1):63–76. https://doi.org/10.1006/exnr.2001.7827.
15. Whalley K, O'Neill P, Ferretti P. Changes in response to spinal cord injury with development: vascularization, hemorrhage and apoptosis. Neuroscience. 2006;137(3):821–32. https://doi.org/10.1016/j.neuroscience.2005.07.064.
16. Fairholm DJ, Turnbull IM. Microangiographic study of experimental spinal cord injuries. J Neurosurg. 1971;35(3):277–86.
17. Figley SA, Khosravi R, Legasto JM, Tseng YF, Fehlings MG. Characterization of vascular disruption and blood-spinal cord barrier permeability following traumatic spinal cord injury. J Neurotrauma. 2014;31(6):541–52. https://doi.org/10.1089/neu.2013.3034.
18. Hsu JY, McKeon R, Goussev S, Werb Z, Lee JU, Trivedi A, Noble-Haeusslein LJ. Matrix metalloproteinase-2 facilitates wound healing events that promote functional recovery after spinal cord injury. J Neurosci. 2006;26(39):9841–50. https://doi.org/10.1523/jneurosci.1993-06.2006.
19. Noble LJ, Donovan F, Igarashi T, Goussev S, Werb Z. Matrix metalloproteinases limit functional recovery after spinal cord injury by modulation of early vascular events. J Neurosci. 2002;22(17):7526–35.
20. Schnell L, Fearn S, Klassen H, Schwab ME, Perry VH. Acute inflammatory responses to mechanical lesions in the CNS: differences between brain and spinal cord. Eur J Neurosci. 1999;11(10):3648–58.
21. Donnelly DJ, Popovich PG. Inflammation and its role in neuroprotection, axonal regeneration and functional recovery after spinal cord injury. Exp Neurol. 2008;209(2):378–88. https://doi.org/10.1016/j.expneurol.2007.06.009.
22. Reese TS, Karnovsky MJ. Fine structural localization of a blood-brain barrier to exogenous peroxidase. J Cell Biol. 1967;34(1):207–17.
23. Imperato-Kalmar EL, McKinney RA, Schnell L, Rubin BP, Schwab ME. Local changes in vascular architecture following partial spinal cord lesion in the rat. Exp Neurol. 1997;145(2 Pt 1):322–8. https://doi.org/10.1006/exnr.1997.6449.
24. Han S, Arnold SA, Sithu SD, Mahoney ET, Geralds JT, Tran P, Benton RL, Maddie MA, D'Souza SE, Whittemore SR, Hagg T. Rescuing vasculature with intravenous angiopoietin-1 and alpha v beta 3 integrin peptide is protective after spinal cord injury. Brain. 2010;133(Pt 4):1026–42. https://doi.org/10.1093/brain/awq034.
25. Choi UH, Ha Y, Huang X, Park SR, Chung J, Hyun DK, Park H, Park HC, Kim SW, Lee M. Hypoxia-inducible expression of vascular endothelial growth factor for the treatment of spinal cord injury in a rat model. J Neurosurg Spine. 2007;7(1):54–60. https://doi.org/10.3171/spi-07/07/054.
26. Herrera JJ, Sundberg LM, Zentilin L, Giacca M, Narayana PA. Sustained expression of vascular endothelial growth factor and angiopoietin-1 improves blood-spinal cord barrier integrity and functional recovery after spinal cord injury. J Neurotrauma. 2010;27(11):2067–76. https://doi.org/10.1089/neu.2010.1403.
27. des Rieux A, De Berdt P, Ansorena E, Ucakar B, Damien J, Schakman O, Audouard E, Bouzin C, Auhl D, Simón-Yarza T, Feron O, Blanco-Prieto MJ, Carmeliet P, Bailly C, Clotman F, Preat V. Vascular endothelial growth factor-loaded injectable hydrogel enhances plasticity in the injured spinal cord. J Biomed Mater Res A. 2014;102(7):2345–55. https://doi.org/10.1002/jbm.a.34915.
28. Wang L, Shi Q, Dai J, Gu Y, Feng Y, Chen L. Increased vascularization promotes functional recovery in the transected spinal cord rats by implanted vascular endothelial growth factor-

targeting collagen scaffold. J Orthop Res. 2018;36(3):1024–34. https://doi.org/10.1002/jor.23678.
29. Widenfalk J, Lipson A, Jubran M, Hofstetter C, Ebendal T, Cao Y, Olson L. Vascular endothelial growth factor improves functional outcome and decreases secondary degeneration in experimental spinal cord contusion injury. Neuroscience. 2003;120(4):951–60.
30. Benton RL, Whittemore SR. VEGF165 therapy exacerbates secondary damage following spinal cord injury. Neurochem Res. 2003;28(11):1693–703.
31. Anderson DK, Saunders RD, Demediuk P, Dugan LL, Braughler JM, Hall ED, Means ED, Horrocks LA. Lipid hydrolysis and peroxidation in injured spinal cord: partial protection with methylprednisolone or vitamin E and selenium. Cent Nerv Syst Trauma. 1985;2(4):257–67.
32. Slotkin JR, Pritchard CD, Luque B, Ye J, Layer RT, Lawrence MS, O'Shea TM, Roy RR, Zhong H, Vollenweider I, Edgerton VR, Courtine G, Woodard EJ, Langer R. Biodegradable scaffolds promote tissue remodeling and functional improvement in non-human primates with acute spinal cord injury. Biomaterials. 2017;123:63–76. https://doi.org/10.1016/j.biomaterials.2017.01.024.
33. Rauch MF, Hynes SR, Bertram J, Redmond A, Robinson R, Williams C, Xu H, Madri JA, Lavik EB. Engineering angiogenesis following spinal cord injury: a coculture of neural progenitor and endothelial cells in a degradable polymer implant leads to an increase in vessel density and formation of the blood-spinal cord barrier. Eur J Neurosci. 2009;29(1):132–45. https://doi.org/10.1111/j.1460-9568.2008.06567.x.
34. Devolder R, Antoniadou E, Kong H. Enzymatically cross-linked injectable alginate-g-pyrrole hydrogels for neovascularization. J Control Release. 2013;172(1):30–7. https://doi.org/10.1016/j.jconrel.2013.07.010.
35. De Laporte L, des Rieux A, Tuinstra HM, Zelivyanskaya ML, De Clerck NM, Postnov AA, Préat V, Shea LD. Vascular endothelial growth factor and fibroblast growth factor 2 delivery from spinal cord bridges to enhance angiogenesis following injury. J Biomed Mater Res A. 2011;98(3):372–82. https://doi.org/10.1002/jbm.a.33112.
36. Li X, Han J, Zhao Y, Ding W, Wei J, Li J, Han S, Shang X, Wang B, Chen B, Xiao Z, Dai J. Functionalized collagen scaffold implantation and cAMP administration collectively facilitate spinal cord regeneration. Acta Biomater. 2016;30:233–45. https://doi.org/10.1016/j.actbio.2015.11.023.
37. Nishida T, Shimokawa H, Oi K, Tatewaki H, Uwatoku T, Abe K, Matsumoto Y, Kajihara N, Eto M, Matsuda T, Yasui H, Takeshita A, Sunagawa K. Extracorporeal cardiac shock wave therapy markedly ameliorates ischemia-induced myocardial dysfunction in pigs in vivo. Circulation. 2004;110(19):3055–61.
38. Lazare JN, Saltzman B, Sotolongo J. Extracorporeal shock wave lithotripsy treatment of spinal cord injury patients. J Urol. 1988;140(2):266–9.
39. Rompe JD, Hope C, Kullmer K, Heine J, Burger R. Analgesic effect of extracorporeal shockwave therapy on chronic tennis elbow. J Bone Joint Surg Br. 1996;78(2):233–7.
40. Wang CJ. Extracorporeal shockwave therapy in musculoskeletal disorders. J Orthop Surg Res. 2012;7:11. https://doi.org/10.1186/1749-799x-7-11.
41. Yamaya S, Ozawa H, Kanno H, Kishimoto KN, Sekiguchi A, Tateda S, Yahata K, Ito K, Shimokawa H, Itoi E. Low-energy extracorporeal shock wave therapy promotes vascular endothelial growth factor expression and improves locomotor recovery after spinal cord injury. J Neurosurg. 2014;121(6):1514–25. https://doi.org/10.3171/2014.8.Jns132562.
42. Reznik JE, Biros E, Sacher Y, Kibrik O, Milanese S, Gordon S, Galea MP. A preliminary investigation on the effect of extracorporeal shock wave therapy as a treatment for neurogenic heterotopic ossification following traumatic brain injury. Part II: effects on function. Brain Inj. 2017;31(4):533–41. https://doi.org/10.1080/02699052.2017.1283060.
43. Reznik JE, Biros E, Lamont AC, Sacher Y, Kibrik O, Milanese S, Gordon S, Galea MP. A preliminary investigation on the effect of extracorporeal shock wave therapy as a treatment for neurogenic heterotopic ossification following traumatic brain injury. Part I: effects on pain. Brain Inj. 2017;31(4):526–32. https://doi.org/10.1080/02699052.2017.1283059.

44. Kang N, Zhang J, Yu X, Ma Y. Radial extracorporeal shock wave therapy improves cerebral blood flow and neurological function in a rat model of cerebral ischemia. Am J Transl Res. 2017;9(4):2000–12.
45. Yu CG, Jimenez O, Marcillo AE, Weider B, Bangerter K, Dietrich WD, Castro S, Yezierski RP. Beneficial effects of modest systemic hypothermia on locomotor function and histopathological damage following contusion-induced spinal cord injury in rats. J Neurosurg. 2000;93(1 Suppl):85–93.
46. Lo TP Jr, Cho KS, Garg MS, Lynch MP, Marcillo AE, Koivisto DL, Stagg M, Abril RM, Patel S, Dietrich WD, Pearse DD. Systemic hypothermia improves histological and functional outcome after cervical spinal cord contusion in rats. J Comp Neurol. 2009;514(5):433–48. https://doi.org/10.1002/cne.22014doi:10.1161/01.Cir.0000148849.51177.97.
47. Kao CH, Chio CC, Lin MT, Yeh CH. Body cooling ameliorating spinal cord injury may be neurogenesis-, anti-inflammation- and angiogenesis-associated in rats. J Trauma. 2011;70(4):885–93. https://doi.org/10.1097/TA.0b013e3181e7456d.
48. Wang D, Yang Z, Zhang J. Treatment of spinal cord injury by mild hypothermia combined with bone marrow mesenchymal stem cells transplantation in rats. Zhongguo Xiu Fu Chong Jian Wai Ke Za Zhi. 2010;24(7):801–5.
49. Wang J, Pearse DD. Therapeutic hypothermia in spinal cord injury: the status of its use and open questions. Int J Mol Sci. 2015;16(8):16848–79. https://doi.org/10.3390/ijms160816848.

Index

A
Adeno-associated virus (AAV), 119
Adipose-derived mesenchymal stem cells (ADSCs), 71
Adipose-derived stem cells (ADSCs), 46, 47
Aminopropyltriethoxysilane (APTES), 48
Amyloid-beta peptides, 8
Amyotrophic lateral sclerosis (ALS), 6
Ankyrin repeat domain 1 (ANKRD1), 39
Anti-A2B5, 3
Anti-GFAP, 3
Astrocyte alignment, 13
Astrocytes
   cell markers, 3
   description, 1
   functions, in CNS
      excess glutamate, removal of, 4
      extracellular homeostasis, control of, 3
      gliotransmission, 6
      glutamatergic neurotransmission, 4
      local blood flow and metabolic support, 4
      signaling, in glial syncytia, 5
      synaptogenesis, 4
      tripartite synapse, 5
   GFAP, 2
   glial cells, 2
   hydrogels, 12
   neural degeneration
      acute lesions, 6
      ALS, 7
      chronic neurodegenerative, 6
      oxidative stress, 6, 7
      reactive astrogliosis, 6
   in neural injury
      amyloid-beta peptides, 8

      BSB, 8
      cellular calcium homeostasis, 7
      glial scar, 7, 8
      neural regeneration, 7
      oxidative stresses, 7
      pro-inflammatory cytokines, 7
      striking changes, 7
      TNFα, 7
      type 1 astrocytes, 8
      type 2 astrocytes, 8
   in neural regeneration, 8, 9
   transplantation (*see* Transplantation, astrocytes)
   types, 2, 3
Atomic force microscopy (AFM), 3
Autologous nerve grafting, 40, 46
Axonal myelination
   deficiencies, 20
   growth factors, 20
   myelin sheath, 20
   OPCs, 20
   propagation rate, 19
Axonal regeneration
   ADSCs, 71
   ESCs, 60, 61, 66
   iPSCs, 68, 69
   MSCs, 69–71
   NSCs, 66–68
Axon demyelination
   electrical stimulation, 28–30
   myelination process, 20
   post-SCI, restoration of, 22, 23
   transplantation, OPCs (*see* OPC transplatation)

B

Basic fibroblast growth factor (bFGF), 46
Bergmann glial cells, 5
Biomaterials, 11
Biomaterial scaffolds, 48, 51, 72
　alginate-fibrinogen hydrogel, 121
　cellular environment characteristics, 120
　collagen filaments function, 72
　DRG, 26
　effective carrier, 71
　ESCs with MSCs, 72
　gels, sponges and conduits, 72
　microenvironment, 25
　microspheres, 26, 27
　nanofibers, 26–28
　nanoparticles, 73
　neural regeneration, 25
　NGF, 26
　novel protocols, 75
　PLG/PLGA, 120
　porous scaffolds, 120
　survival rate, grafted neural stem cells, 73
　3D printing technology, 74
　use of, 120
　VEGF, 26
　vehicles, 73
Bipotential progenitor cells, 2
Blood-brain barrier (BBB), 8
Blood spinal barrier (BSB), 8
Bone marrow-derived MSCs (BMSCs), 70, 72
Brain-derived neurotrophic factor (BDNF), 44, 87

C

Calcium signaling, 96
Cell division, EFs, 92, 94
Cell polarization, 93
Central nervous system (CNS), 38, 60, 61
　astrocytes (*see* Astrocytes)
　components, 1
　SVZ, 22
　transplantation, wild-type astrocytes, 11
Cerebrospinal fluid (CSF), 114
Chain migration, 88
Chinese hamster ovary (CHO) cells, 93, 95, 98
Cotransplantation, SCs, 51
Cyclin-dependent kinase 5 (CDK5), 10, 11

D

Diblock copolypeptide hydrogel (DCH), 73
Direct-current electric field (DC EF), 86
Dorsal root ganglia (DRG), 26

E

EF-guided migration
　and calcium signals growth cone navigation, 96
　cell membrane receptors, 97, 98
　cell migration from oligospheres, 91
　cell polarization, 93–95
　integrins, 97
　intracellular signaling pathways, 96, 97, 99
　neurogenesis, 89
　neuron migration, 89
　stem cells and stem cell-derived neural cells, 90, 91
Electric field (EF)
　axonal growth, 85, 86
　cell division and migration, 92, 94
　chemotaxis and electrotaxis, 86
　in CNS development and regenerating tissues, 99, 100
　to guide cell migration, in neurogenesis, 101
　guided migration (*see* EF-guided migration)
　nerve and spinal cord regeneration *in vivo*, 100, 101
　neuron migration, 87–89
　oriented cell division, 92, 93
Electrospinning techniques, 49
Embryoid bodies (EBs), 75
Embryonic stem cells (ESCs)
　description, 60
　EF-guided migration, 90
　graft, 61
　hESC, 61
　human embryonic neural progenitors, 61
　human ESC-derived OPCs, transplantation, 61
　neural development, 66
　neurospheres, 61
　neurotrophic factors, 61
Excitatory amino acid transporters (EAATs), 4
Extracellular homeostasis, 3
Extracorporeal shock wave therapy (ESWT)
　in adult female Sprague-Dawley rat models, 121
　functional (fESWT), 121
　and hypothermia, 121
　radial (rEWST), 122
　shock wave, 121
　VEGF, 122

Index

F
Fibroblast growth factor (FGF), 20

G
Galvanotaxis, 88
Gene therapy, 45
Glial cell line-derived neurotrophic factor (GDNF), 45
Glial fibrillary acidic protein (GFAP), 2
Glial-restricted precursor (GRP), 9
Gliophilic migration, 88
Growth factors
  collagen, 39, 46
  fibronectin, 39, 50
  GH, 44
  heparin sulfate, 39
  laminin, 39, 46, 50
Growth hormone (GH), 44
Guanosine triphosphate (GTP)ases, 86, 94

H
Histone deacetylase 3 (HDAC3), 39
Human embryonic stem cell (hESC), 61, 62, 68, 69
Human embryonic stem cell-derived mesenchymal stem cells (hESC-MSCs), 45
Human iPSC (hiPSC), 64, 69, 74, 75
Human telomerase reverse transcriptase gene (hTERT), 51
Human umbilical mesenchymal stem cells (HUMSCs), 70
Hyaluronic acid, 72
Hydrogels, 49
Hypothermia
  bone marrow mesenchymal stem cell transplantation, 122
  corticosteroids, use of, 112
  ESWT, 121
  profound, moderate and modest, 122
  in rat compression SCI model, 122
  revascularization, 123

I
Induced pluripotent stem cells (iPSCs)
  axon regeneration, 68
  description, 68
  vs. ESCs, 68
  iPSC-derived neurospheres, 69
  NSPCs, 69
  primary skin fibroblasts, 68

Integrins, 97, 98
iPSC-derived neurospheres, 69
Ischemia, 112, 113, 121

K
Kruppel-like factor 7 (KLF7), 45

M
Maghemite nanoparticles, 73
Matrix metalloproteinases (MMP), 114
Mechanosensitivity, 39
Mesenchymal stem cells (MSCs)
  ADSCs, 71
  BMSCs, 70
  description, 69
  EF-guided migration, 91
  human umbilical cord-derived MSCs, 70
  HUMSCs, 70
  use of, 70
Metformine, 44
Methylprednisolone sodium succinate (MPSS), 118, 119
Micro-anatomical domains, 2
Microenvironment, 22, 25
Microspheres, 11
Mitogen-activated protein kinase (MAPK), 30, 98
Molecular therapy, 112, 117
Motor-evoked potential (MEP), 24
Myelination, SCs, 50, 51

N
Nanofibers, 49, 50
Nano, micro and nanofiber micropore (NFMP), 9
Nanoparticles, 73
Nerve growth factor (NGF), 12, 46, 49, 50
Nerve injury, 38–40
Neural crest cells (NCCs), 37, 38
Neural differentiation, 73
Neural progenitor cells (NPCs), 90
Neural regeneration
  SCs and biomaterial, 46, 48–50
  stem cell-derived SCs, 45, 46
Neural stem cells (NSCs)
  description, 66
  embryonically derived, 68
  ependymal cells, 66
  human fetal NSPCs, transplantation, 67
  oligodendrocytes, 67

Neural stem progenitor cells (NSPCs), 67, 69, 72, 97, 98
Neurogenesis, 85, 89, 92, 101
Neurotransmitters, 3–5, 9, 10
Neurovascular unit, 2
Nitric oxide (NO), 21

O
Olfactory ensheathing glia cells (OEGCs), 51
Olfactory unsheathing cells (OECs), 25
Oligodendrocyte (OL), 19
Oligodendrocyte precursor cells (OPCs), 20, 89, 90
   abluminal endothelial surface, 22
   chromodomain helicase DNA, 23
   embryonic development, 20
   growth factors, 23
   migration, 22
   myelin sheath, 20
   PAR2 knockout mice model, 23
   proliferative response, 23
   remyelinating damaged axons, 23
OPC transplatation
   Basso–Beattie–Bresnahan locomotor, 25
   CNS lesion, 24
   hind limb locomotor function, 24
   MEP, 24
   necrosis factor, 24
   OECs, 25
   regulated miRNAs, 24
   remyelination, 24
Oxidative stress, 6, 7

P
PAR2 knockout mice model, 23
Parvin, 99
Paxillin, 99
Peripheral nerve
   BDNF, 44
   and cultured *in vitro*, 50
   frankincense extract, 44
   functional recovery, 45
   injury, 38, 39
   repair, 40
   sural nerve biopsy, 40
Platelet-derived growth factor (PDGF), 20
Polarity, 90, 94, 95, 98
Poly-lactic-co-glycolic acid (PLG/PLGA), 44, 50, 120
Poly-L-lysine (PLL) scaffolds, 120

R
Radial ESWT (rEWST), 122
Reactive astrogliosis, 6
Reconstruction, vascular structure, 118, 119
   biomaterial scaffolds, 120, 121
   ESWT, 121, 122
   hypothermia, 122, 123
   therapeutic molecules, 116, 119
Revascularization, 113–115
RNA sequencing, 98, 99

S
Scaffold-free approach, 13
Schwann cell precursors (SCPs), 37, 38
Schwann cells (SCs), 22
   autologous nerve grafting, 40
   and biomaterial
      description, 47–48
      hydrogel, in neural repair, 49
      nanofibers, 49
      scaffolds, 46, 48
      synthetic and natural, 46
      tissue-engineered nerve graft, 46
   denervated, 40
   gene therapy, 45
   grafting, 41–43
   growth factors, 44
   myelinating nerves, 38
   NCCs, 37
   nerve injury, 38
   origin, 37
   peripheral nerve injury, 39
   SCI (*see* Spinal cord injury (SCI))
   SCPs, 37
   SOX-10, 38
   stem cell-derived, 45, 46
   synthetic and biological molecules, 44
   zinc-finger protein ZEB2, 38
Sciatic nerve injury, 48, 49
Signaling pathways, EF-guided cell migration, 98, 99
Spinal cord
   extrinsic arteries, 112
   intrinsic arteries, 112
   venous system, 112
Spinal cord injury (SCI)
   application, 60
   axonal damage (*see* Axonal regeneration, SCI)
   axon demyelination, 20
   caspases function, 21

mitochondrial fragmentation, 21
OL apoptosis, 21
OL loss, 20
biomaterial scaffolds, 51
cascade reaction, 21
causes, 50
CNS, 111
co-transplatation (*see* Biomaterial scaffolds)
EF-guided migration (*see* EF-guided migration)
myelination potential, SCs, 51
nervous system, 60
peripheral arterial supply, 112
regeneration, 51
remyelination, SCs, 50
SC myelination, 50
SC transplantation, 50, 51
SCs efficacy, 52
stem cells, transplantation of, 62–66
transplanted stem cells, 60
traumatic and atraumatic, 111
traumatic SCI, 59
vascularization (*see* Vascularization, in spinal cord)
vascular and neural networks, 112
vascular rupture, 20
Stem cell, EF-guided migration, 90, 91
Striking changes, 7
Subventricular zone (SVZ), 22
Superoxide dismutase (SOD1) enzyme, 7
Synaptic terminals, 5
Synaptogenic signals, 4

T
Therapeutic molecules, 116
Time-lapse imaging system, 11
Tissue-engineered nerve graft (TENG), 46
Transplantation, 22
    astrocytes
        ALS, 11
        biomaterials, 11
        CDK5, 10, 11
        cerebral ischemia, 10
        CNS stem cell-based therapies, 14
        gene expression, 13
        glial cells, 10
        GLT-1, 11
        glutamate transporter, 11
        GRPs, 11
        microspheres, 11
        neural injury, 10
        neural precursor cells, 10
        neural tissue lesion, 11
        NGF, 12
        NSCs, 10
        regenerated fibers, 10
        scaffold-free approach, 13
        scar tissue formation, 10
        stem cell, 10
        stroke recovery, 10
        therapeutic agents, 11
    and biomaterials, 71
    BMSCs, 70
    ependymal stem progenitor cells, 67
    ESC-derived neural aggregates, 61
    ESC-derived neurospheres, 61
    ESC-derived OPCs, 60
    iPSC-derived NSCs, 68
    iPSCs, 75
    MSCs, 69
    murine iPSC-derived astrocytes, 69
    NSPCs, 67
    peripheral nerve graft, 60
    pre-differentiated ESCs, 66
    SCs, 45, 50, 51
    spatiotemporal imaging, 73
    stem cells, SCI, 62–66
Tripartite synapse, 5

V
Vascular endothelial growth factor (VEGF), 26, 119–122
Vascularization, in spinal cord
    atraumatic causes, 113
    damage, 113, 114
    MMP, 114
    natural process, revascularization, 114, 115
    primary injury, 113
    reconstruction (*see* Reconstruction, vascular structure)
    revascularization, 113
    secondary injury, 113
Voltage-gated Ca^{2+} channels (VGCCs), 96
Voltage-gated sodium (Na) channel (VGNC), 96

Z
Zinc-finger protein (ZEB2), 38